Rapid Assessment Program

RAP Bulletin of Biological Assessment

A Marine Rapid Assessment of the Togean and Banggai Islands, Sulawesi, Indonesia

Gerald R. Allen and Sheila A. McKenna
Editors

Center for Applied Biodiversity Science (CABS)

Conservation International

Hasanuddin University

Australia Institute of Marine Science

Western Australia Museum

The RAP Bulletin of Biological Assessment is published by:
Conservation International
Center for Applied Biodiversity Science
Department of Conservation Biology
1919 M St. NW, Suite 600
Washington, DC 20036
USA
202-912-1000 telephone
202-912-1030 fax
www.conservation.org
www.biodiversityscience.org

Editors: Gerald R. Allen, and Sheila A. McKenna
Design/Production: Kim Meek, Conservation International
Production Assistant: Fabian Painemilla
Maps: M. Denil and Vineet Katariya
Cover Photographs: Gerald R. Allen

RAP Bulletin of Biological Assessment was formerly RAP Working Papers. Numbers 1–13 of this series were published under the previous title.

Suggested citation:
Allen, G. R. and S. A. McKenna (eds.). 2001.
A Marine Rapid Assessment of the Togean and Banggai Islands, Sulawesi, Indonesia. RAP Bulletin of Biological Assessment 20, Conservation International, Washington, DC.

Printed on recycled paper (80% recycled/60% post-consumer waste)

Using New Leaf Opaque 60# smooth text paper (80% recycled/60% post-consumer waste), and bleached without the use of chlorine or chlorine compounds results in measurable environmental benefits[1]. For this report, using 1,404 pounds of post-consumer waste instead of virgin fiber saved...

8	Trees
763	Pounds of solid waste
840	Gallons of water
1,095	Kilowatt hours of electricity (equal to 1.4 months of electric power required by the average U.S. home)
1,387	Pounds of greenhouse gases (equal to 1,123 miles travelled in the average American car)
6	Pounds of HAPs, VOCs, and AOX combined
2	Cubic yards of landfill space

[1] Environmental benefits are calculated based on research done by the Environmental Defense Fund and the other members of the Paper Task Force who studied the environmental impacts of the paper industry. Contact the EDF for a copy of their report and the latest updates on their data. Trees saved calculation based on trees with a 10" diameter. Actual diameter of trees cut for pulp range from 6" up to very large, old growth trees. Home energy use equivalent provided by Pacific Gas and Electric Co., San Francisco. Hazardous Air Pollutants (HAPs), Volatile Organic Compounds (VOCs), and Absorbable Organic Compounds (AOX). Landfill space saved based on American Paper Institute, Inc. publication, Paper Recycling and its Role in Solid Waste Management.

Table of Contents

Participants

Gerald R. Allen, Ph.D. (Ichthyology and
Science Team Leader)
Center for Applied Biodiversity Science
Conservation International
1919 M Street NW, Suite 600
Washington, DC 20036
USA

Mailing address:
1 Dreyer Road
Roleystone, WA 6111
Australia
Fax: (618) 9397 6985
Email: tropical_reef@bigpond.com

Khaerul Anwar (Marine Resources)
Sekber Konsorsium
Togean, Palu, Central Sulawesi
Indonesia

Douglas Fenner, Ph.D. (Corals)
Australian Institute of Marine Science
P.M.B No. 3
Townsville, Queensland 4810
Australia
Email: d.fenner@aims.gov.au

La Tanda (Reef Fisheries)
Balai Penelitian dan Pengembangan Sumberdaya Laut
Puslitbang Oseanologi LIPI Biak
Biak, Irian Jaya
Indonesia

Purbasari Surjadi (Marine Resources)
Conservation International-Indonesia
Jl. Taman Margasatwa No. 61
Pasar Minggu, Jakarta 12540
Indonesia
Fax: (62) (21) 780 1265
Email: ci-indonesia@conservation.org

Fred E. Wells, Ph.D. (Malacology)
Department of Aquatic Zoology
Western Australian Museum
Francis Street
Perth, WA 6000
Australia
Email: wellsf@museum.wa.gov.au

Timothy Werner, M.Sc. (RAP Survey Team Leader)
Center for Applied Biodiversity Science
Conservation International
1919 M St. NW, Suite 600
Washington, DC 20036
USA
Fax: (1) (202) 912 1030
Email: t.werner@conservation.org

Syafyudin Yusuf (Reef Ecology)
Marine Biology Laboratory
Hasanuddin University
Macassar, South Sulawesi
Indonesia
Email: syafyudin.yusuf@usa.net

Organizational Profiles

Conservation International

Conservation International (CI) is an international, non-profit organization based in Washington, DC. CI acts on the belief that the Earth's natural heritage must be maintained if future generations are to thrive spiritually, culturally, and economically. Our mission is to conserve biological diversity and the ecological processes that support life on earth, and to demonstrate that human societies are able to live harmoniously with nature.

Conservation International
1919 M Street NW, Suite 600
Washington, DC 20036 USA
(1) (202) 912-1000 telephone
(1) (202) 912-1030 fax
http://www.conservation.org

Australian Institute of Marine Science

The mission of the Australian Institute of Marine Science (AIMS) is to generate the knowledge to support the sustainable use and protection of the marine environment through innovative, world-class scientific and technological research. It is an Australian Commonwealth Statutory Authority established by the Australian Institute of Marine Science Act of 1972 in recognition of a national need to manage Australia's marine environment and marine resources.

Australian Institute of Marine Science
Cape Ferguson, Queensland
PMB No 3, Townsville MC QLD 4810
Australia
(07) 4753 4444 telephone
(07) 4772 5852 (fax)
http://www.aims.gov.au

Hasanuddin University

As an institution for education, research, and development, Hasanuddin University (UNHAS) aims to strengthen and promote science, technology, and arts that are needed for improvement in the quality of human welfare, especially that of the Indonesian people.

Hasanuddin University
Tamalanrea Campus
Jl. Perintis Kemerdekaan km 10
Makassar - Indonesia 90245
Phone: 062 0411 584002
Fax: 062 0411 585188
Email: rektorat@unhas.ac.id

Western Australian Museum

The Western Australian Museum was established in 1891 and its initial collections were of geological, ethnological and biological specimens. The 1960s and 1970s saw the addition of responsibility to develop and maintain the State's anthropological, archaeological, maritime archaeological and social and cultural history collections. The collections, currently numbering over two million specimens/artifacts, are the primary focus of research by the Museum's staff and others. The aim is to advance knowledge on them and communicate it to the public through a variety of media, but particularly through a program of exhibitions and publications.

Western Australian Museum
Francis Street
Perth, WA 6000
Australia
(08) 9427 2716 (telephone)
(08) 9328 8686
http://www.museum.wa.gov.au

Acknowledgments

The survey and report would not have been possible without the support and guidance of CI-Indonesia, particularly Jatna Supriatna, Purbasari Surjadi, Suer Surjadi, Myrna Kusuma-wardhani, and Ermayanti. We are also grateful for the support of Lembaga Ilmu Pengetahuan Indonesia (LIPI), especially Soehartono Soedargo, the Director of the Bureau of Science and Technical Cooperation, and his capable assistant Ina Syarief. We are also indebted to H. Ono Kurnaen Sumadhiharga, the Director of Pusat Penelitian dan Pengembangan Oseanologi, for sponsorship of the RAP survey. For their help on the report's map, we thank Vineet Katariya and Mark Denil.

We also extend our sincere thanks to owner Hanny Batuna, and the crew of the Serenade, for providing an excellent base of operation for the survey. We also thank Angelique Batuna for her assistance in arranging the boat charter and ground transport.

The Rufford Foundation and the Henry Foundation generously provided financial support for the project.

Executive Summary

Introduction

This report presents the results of a rapid field assessment of the rich and previously undocumented marine biodiversity found in the Togean and Banggai Islands, Sulawesi, Indonesia. The area lies near the center of global marine biodiversity or "coral triangle," composed of Indonesia, Philippines, Malaysia, Papua New Guinea, Japan, and Australia. This region harbors the most biologically diverse coral reefs, mangroves, and seagrass beds in the world. Although the Indonesian-Philippines region is widely acclaimed as the richest area within the triangle, there is insufficient detailed information to support this claim.

This survey was implemented by the Marine Rapid Assessment Program (RAP) of the Center for Applied Biodiversity Science at Conservation International (CI) in collaboration with the Indonesian Institute of Sciences (LIPI). It represents a vital step in CI's integrated biological surveys of the "coral triangle," with the express aims of assessing biodiversity in the region and providing guidelines for its conservation.

Overview of Marine RAP

The goal of Marine RAP is to rapidly generate and disseminate information on coastal and near-shore shallow-water marine biodiversity for conservation purposes, with a particular focus on recommending priorities for conservation area establishment and management. Marine RAP deploys multi-disciplinary teams of marine scientists and coastal resource experts to determine the biodiversity significance and conservation opportunities of selected areas. This is accomplished through underwater inventories that generally last about three weeks. Marine RAP surveys produce species lists that serve as indicators of overall biological richness, as well as recording several measurements to assess overall ecosystem health. During each survey, RAP supports parallel assessments of local human community needs and concerns, which become incorporated into the final recommendations.

By comparing the results obtained from many surveys, Marine RAP is ultimately focused on ensuring that a representative sample of marine biodiversity is conserved within protected areas and through other conservation measures.

Togean and Banggai Islands

The Togean Islands stretch for a distance of 90 km across the center of Tomini Bay in northern Sulawesi. The archipelago is composed of six main, hilly (to 543-m elevation) islands and 60 relatively low, satellite islands. The westernmost islands of Batudaka and Togean are the largest in the archipelago, spanning a combined longitudinal distance of about 50 km, and separated by a narrow channel. The population of the group is about 30,000. The area supports a rich diversity of coral reef habitats, including fringing reefs, barrier reefs, and atolls. The majority of the population depends on marine resources for food and income.

The Banggai Islands are situated about 115 km south of the Togeans in the northern section of Tolo Bay. Peleng, an 80 km-long mountainous island, occupies most of the land area. The remainder of the archipelago consists of three mountainous islands (Banggai, Labobo, and Bangkulu) surrounded by narrow fringing reefs, and a host of satellite islands, cays, and shoals. The population of 91,000 is mainly concentrated on the larger islands. As in the Togeans, poverty is widespread and the population depends on farming and fishing for sustenance and capital. There are no formal conservation programs in existence and the sustainability of natural resources is a primary concern.

The Togean-Banggai Survey

The Marine RAP survey of the Togean and Banggai Islands assessed 47 sites over a 17-day period (October 24–November 9, 1998). Sites were pre-selected to maximize the diversity of habitats surveyed, and thus produce a list of species for the area that was as comprehensive as possible. At each site, an underwater inventory was made of three faunal groups selected to serve as indicators of overall coral reef biodiversity: reef corals, molluscs, and fishes. Additional observations were made on the environmental condition of each site, including the presence and abundance of species exploited for local consumption or commercial sale.

Summary of Results

The Togean and Banggai Islands exhibit rich marine biodiversity, but there is concern about over-exploitation of marine resources and the widespread use of illegal fishing methods. Some of the most significant findings of the Marine RAP survey include:

- The Togean and Banggai Islands have a highly diverse coral fauna containing at least 314 species. At least six new coral species and numerous new records for Indonesia were documented during the RAP survey.

- The percentage of live coral cover for survey sites averaged 41-42%. A total of 13 sites had an average in excess of 50%. These values compare favorably with other surveyed areas in Indonesia.

- Coral bleaching was observed at nearly every site in the Togean Islands, but at only one site in the Banggai Islands. Widespread bleaching in the Togeans appears to be correlated with increased sea temperatures, which ranged from 30–33°C.

- A total of 541 species of molluscs belonging to 103 families were recorded. These figures are slightly lower than previous CI surveys at Milne Bay (Papua New Guinea) and the Calamianes Islands (Philippines). However, our mollusc expert was only able to participate in the first half of the RAP survey, and thus the Banggai Islands were not sampled.

- The Togean and Banggai Islands have a diverse reef fish fauna. A total of 819 species were observed or collected during the present survey. An extrapolation method utilizing six key index families indicates a total fauna consisting of at least 1,023 species.

- A total of 147 edible (commercial) species in 38 genera were classified as target fishes. The reef fish biomass in the area was estimated at 5.33–298.27 ton/km^2.

- Illegal fishing practices, including the use of dynamite, cyanide, and various types of netting are common in both island groups. Dynamite damage was observed at 86% of sites.

- Encouraging signs for conservation action include the formation of Sekber Konsorsium Togean, a community-based organization involved with the development of an integrated system of marine and terrestrial reserves. This organization consists of coastal communities, local government, and other stakeholders working together to achieve consensus on matters relating to the designation, delineation, and management of protected areas.

Conservation Recommendations

The results of the Marine RAP survey indicate that the reefs of the Togean-Banggai area harbor a remarkably diverse biological community. Although they are by no means pristine, many reefs feature extensive live coral cover and minimal damage from humans. Conversely, there are also numerous areas severely affected by the widespread use of explosives and cyanide, as well as by poor agricultural practices that literally choke local reefs with layers of silt. It is therefore crucial to implement proper conservation practices before it is too late to save the irreplaceable marine resources of this unique region. Some specific recommendations are as follows:

1) **Evaluate and prevent downstream environmental impacts of land-based activities.** Uncontrolled clearing of native forests for agriculture increases erosion and consequent high levels of sediment in natural watercourses. Once discharged into the sea it is only a matter of time before local reefs are covered with layers of fine silt. Therefore, there is a critical need for watershed protection and a coordinated management effort that considers both terrestrial and marine ecosystems as an integrated system.

2) **Establish marine protected areas.** An increasing population, general shortage of cash, and dwindling marine resources combine to accelerate over-harvesting and more reliance on illegal, destructive fishing methods such as explosives and cyanide. These trends are already well established and, if allowed to continue unabated, the future of local reef communities is bleak. For the benefit of both human and reef communities it is essential to establish an effective network of MPAs that capture a representative cross-section of marine habitats.

3) **Strengthen species conservation programs for rare and endangered marine wildlife.** Indonesia has a comprehensive assortment of legislation for the protection of

wildlife, but all too often there are no effective management plans for their implementation, nor is there any enforcement capability to prevent the illegal harvesting of protected species. There is a genuine need to develop conservation programs aimed at restoring drastically declining populations such as sharks, turtles, dugong, giant clams, and large fish species.

4) **Enact species conservation measures for *Pterapogon kauderni* to include an international ban on the harvest and trade of wild specimens.** The Banggai cardinalfish (*Pterapogon kauderni*) is perhaps the only reef fish in the world whose continued existence is genuinely threatened. It occurs only in the Banggai Islands, has an extremely low fecundity, and is currently harvested in huge numbers for the international aquarium trade. The survival of this species depends on effective conservation action.

5) **Examine feasibility of developing economic incentives that support marine conservation.** The Togean and Banggai Islands offer a diversity of marine habitats, including world-class diving sites. The area has strong potential to support marine-oriented ecotourism. However, more study is needed of the socio-economic factors involved.

6) **Strengthen capacity within the region for managing natural environments.** Local governments typically lack sufficient funding or staffing for conservation programs. The process will most likely need to be encouraged through participation and funding by outside agencies, either national or international. NGOs are particularly good candidates for providing assistance.

7) **Enforce existing laws.** Provincial and national laws governing the protection of wildlife and prohibiting the use of destructive fishing methods such as explosives and cyanide need to be strictly enforced if marine resources are to be sustained. Analyze potential enforcement systems available and implement the one most cost effective. Funds need to be allocated to provide sufficient marine enforcement capacity for local police or other authorized personnel.

8) **Launch an environmental awareness campaign.** Local residents should be aware of the unique nature of their surrounding marine environment and the endemic organisms that live there. Environmental awareness can be achieved in a variety of ways, including primary and secondary school curricula, guest speakers at town meetings, posters, and educational videos.

9) **Establish a long-term environmental monitoring program.** Local communities should be involved with all aspects of their marine resource use and conservation, including participation in environmental monitoring. Simple monitoring protocols need to be introduced to periodically assess the condition of local reefs and associated resources to ensure they remain sustainable.

10) **Promote collection of data essential for marine conservation planning.** Sound management of natural resources is dependent on a wealth of biological and non-biological information. Baseline studies involving taxonomic inventories and basic ecology are important, as are cultural and socio-economic aspects related to resource use.

Overview

Introduction

The Togean and Banggai Islands are situated north and south, respectively, of the prominent, eastward-projecting peninsula of central Sulawesi. The Togean group occupies the central portion of Tomini Bay, stretching over a distance of about 90 km. This small archipelago contains 66 islands of which Una-una, Batudaka, Togean, Talatakoh, Waleakodi, and Waleabahi are the largest. Una-una, a recently active volcano, is relatively isolated, situated about 30 km north of Batudaka. The land area of the Togean Group covers about 755 km^2, mainly consisting of hilly or mountainous terrain. The maximum elevation on the six main islands ranges from 354 to 543 m.

The Banggai Islands lie about 115 km south of the Togeans in the northern section of Tolo Bay. Peleng, with a length of 80 km and widths of up to 45 km, is by far the largest member of the group. It is clearly visible from the mainland city of Luwuk, situated across the 30-km wide Peleng Strait. The remainder of the archipelago consists of three mountainous islands (Banggai, Labobo, and Bangkulu), and a host of satellite islands, cays, and shoals. The islands occupy a land area of 1,912 km^2 and have a maximum elevation of 1,059 m. Population and area statistics for the Banggai and Togean groups are summarized in Table 1.

The area is important for fisheries. According to catch statistics for the mid-1990s, the Seram Sea/Tomini Bay area produces the second highest volume of small pelagic fishes caught in Indonesia (National Coordination Agency for Surveys and Mapping [Bakosurtanal], 1998. Indonesia: Marine Resources Atlas). Both the Togean and Banggai groups are populated by dugong and marine turtles, and these islands are also one of the last bastions in eastern Indonesia for the endangered coconut crab.

The combined Togean-Banggai area supports a rich diversity of coral reefs but, in spite of the relatively low human population density, there are clear indications that at least some marine resources are being over-exploited. This is perhaps understandable, considering that the sea represents the most important source of food and income for the majority of islanders. Indeed, it is estimated that 80% of the population is directly involved with the harvest of marine resources. This report is intended to serve as a guide for future investigations and provide crucial baseline information for the sound management and conservation of coral reefs and their associated fauna.

Table 1. Population and area data for Togean and Banggai Islands. (Source: Regency Statistics)

Regency	Villages	Area (km²)	Population
Togean Islands			
Una-Una	21	515.19	17,690
Walea Kepulauan	16	240.00	11,803
Banggai Islands			
Lo Bangkurung	25	413.28	19,552
Banggai Is.	20	294.39	23,643
Tinangkung	22	593.49	17,588
Totikum	18	294.24	15,943
Liang	24	316.19	14,291

Marine RAP—Rationale and Methodology

The goal of Marine RAP is to rapidly generate and disseminate information on coastal and near-shore shallow-water marine biodiversity for conservation purposes, with a particular focus on recommending priorities for the establishment and management of conservation areas. In addition to recording the diversity of selected taxa which are used as indicators of an area's overall biological diversity, Marine RAP surveys report on the physical nature of survey sites and their environmental conditions. They also support parallel assessments of the social conditions that influence conservation in the survey areas. Marine RAP therefore involves a multi-disciplinary approach to produce appropriate and realistic conservation recommendations based on an analysis of biological data in tandem with social and environmental information.

The information produced by Marine RAP surveys is best applied in raising awareness about the importance of marine conservation, and focusing ecosystem management plans and activities on critical habitats and key issues. It is a layer of information that is generally not available to government officials, local people and others who are responsible for making decisions about the use and management of marine resources. A primary goal in marine conservation is to design and implement policies and projects that can confidently be said to protect a representative sample of marine biodiversity. Without an understanding of where this biodiversity occurs and how it is distributed, it is difficult to confirm that management programs are effective. It is the goal of Marine RAP to provide this critical layer of biological information, which is needed to ensure that marine conservation efforts will ultimately have their intended benefit.

Many other layers of information are important in marine conservation efforts, for example the location of spawning fish aggregations, turtle nesting beaches, culturally valuable sites, and larval source areas. Ideally, the data on biodiversity and environmental condition generated by Marine RAP will be combined with these types of information. However, it is not the goal of Marine RAP to compile all layers of information needed for marine conservation during the course of a rapid biodiversity survey. The focus of Marine RAP is on generating one of the most essential, yet largely missing, layers, which can serve as a foundation for incorporating additional information once it becomes available.

Endemism is often used as a basis for determining relative conservation priority, particularly in terrestrial ecosystems. In general, endemism is much less pronounced in the sea than it is on land. This is particularly true throughout the "coral triangle" (i.e., "Australasia"—the area including northern Australia, the Malay-Indonesian Archipelago, Philippines, and western Melanesia), considered to be the world's richest area for marine biodiversity. The considerable homogeneity found in tropical inshore communities is in large part due to the pelagic larval stage typical of most organisms. For example, reef fish larvae are commonly pelagic for periods ranging from 9–100 days (Leis, 1991). A general lack of physical isolating barriers and numerous island "stepping stones" have facilitated the wide dispersal of larvae throughout the Indo-Pacific.

In the absence of high endemism, overall species richness and relative abundance take on important meaning in conservation priority-setting in marine ecosystems. Extensive biological survey experience over a broad geographic range yields the best results. This enables the observer to recognize any unique assemblages within the community, unusually high numbers of normally rare taxa, or the presence of any unusual environmental features.

Reef corals, fishes, and molluscs are the primary biodiversity indicator groups used in Marine RAP surveys. Corals provide the major environmental framework for fishes and a host of other organisms. Without reef-building corals, there is limited biodiversity. This is dramatically demonstrated in areas consisting primarily of sand, rubble, or seaweeds. Fishes are an excellent survey group as they are the most obvious inhabitants of the reef and account for a large proportion of the reef's overall biomass. Furthermore, fishes depend on a huge variety of plants and invertebrates for their nutrition. Therefore, areas rich in fishes invariably have a wealth of plants and invertebrates. Molluscs represent the largest phylum in the marine environment, the group is relatively well known taxonomically, and they are ecologically and economically important. Mollusc diversity is exceedingly high in the tropical waters of the Indo-Pacific, particularly in coral reef environments. Gosliner et al. (1996) estimated that approximately 60% of all marine invertebrate species in this extensive region are molluscs. Molluscs are particularly useful as a biodiversity indicator for ecosystems adjacent to reefs where corals are generally absent or scarce (e.g., mud, sand, and rubble bottoms).

Historical Notes

Prior to this Marine RAP (October 24–November 9, 1998), no comprehensive faunal surveys of corals, molluscs, or fishes had been undertaken in the combined Togean-Banggai area. Indeed, until the last decade, there appears to have been very little marine biological exploration of any sort. One noteworthy exception was the collection of two specimens of a small cardinalfish in 1920 by a resident Dutch medical officer. These were sent to the Natural History Museum in Leiden, and the species was eventually described by Koumans (1933) as *Pterapogon kauderni*. The recent rediscovery of this exquisite fish has created a sensation in the marine aquarium fish industry and huge numbers are presently harvested from the Banggai Islands, its only known locality.

The only previous surveys include an investigation of corals and fishes of the Banggai Islands by a team from the Indonesian Institute of Sciences' Oceanology Section (PPPO) in 1994–1995 (Suharsono et al., 1995). A total of 181 fishes and 157 corals were recorded during this expedition. In addition, C. Wallace of the Museum of Tropical Queensland (Townsville, Australia) investigated the *Acropora* coral fauna of the Togean Islands in 1995. The published results (Wallace and Wolstenholme, 1998) revealed an unusually rich fauna consisting of 61 species, one of the highest totals recorded anywhere for this genus. Wallace also conducted a marine biodiversity survey of the Togeans with the aid of an international team of biologists in October 1999. The results of this survey were presented at the 9th International Coral Reef Symposium (Wallace, 2000).

Physical Environment

The climate is monsoonal, but rainfall is relatively low compared to many other parts of Indonesia. The heaviest rainfall at Gorontalo, Sulawesi (80 km north of the Togeans) occurs in two periods, between April–July and November–December, when monthly rainfall fluctuates between 110–119 mm per month. At Luwuk, near the Banggai Islands, the wettest (122–184 mm) months occur between March and August. Yearly figures based on average monthly rainfall between 1971 and 1985 for Gorontalo and Luwuk are 1251 and 1162 mm respectively. However, Suharsono et al. (1995) gave an average yearly rainfall of 300–1000 mm for the Banggai Islands.

Predominant winds in Indonesia are the southeast (May–September) and northwest (November–March) monsoons, although the pattern is highly variable depending on locality. Reef morphology at a number of sites visited during the survey indicate that the prevailing winds are generally from the northeast, but their influence on the Togean Islands is moderated by the Minahasa Peninsula. This same sheltering effect is also present on Peleng Island in the Banggais due to its close proximity to the mountainous central peninsula of the eastern Sulawesi mainland.

Prevailing surface currents flow northward past the Banggai Islands and eastern Sulawesi through the Maluku Sea. During the northwest monsoon, these currents are fed by the Banda Sea to the south. In August they are fed mainly by a westerly flow from the Seram Sea (National Coordination Agency for Surveys and Mapping [Bakosurtanal]. 1998. Indonesia: Marine Resources Atlas).

Tidal fluctuations in the area are relatively slight. The maximum spring-tidal fluctuation is only about 1 m. Currents were generally slight or absent at all dive sites except two. There was virtually no current at the beginning of the dive at Cape Dongolalo (site 28), but towards the end severe currents were experienced, estimated to be at least 6 knots. Conversely, a 3 knot current was present at the beginning of the dive at Pontil Kecil Island (site 30), but by the end of the dive the current was negligible.

Sea temperatures during the survey ranged from about 29 to 33°C, and for most sites were consistently in the 30–31°C range. Temperature tended to be slightly cooler in the Banggai Islands compared to the Togeans. The general temperature regime was warmer in comparison to CI surveys in the Philippines and Milne Bay. The correlation between elevated sea temperatures and coral bleaching has been well documented (Lesser, 1996). Significantly, we recorded bleaching at 21 of 24 sites in the Togean Islands where the warmest temperatures were evident. In stark contrast, this phenomenon was noted at only one site in the Banggai Group.

Human Environment

The Togean Islands are administrated by the Poso Regency and are subdivided into two sub-districts (*kecamatan*), with 37 villages (*desa*) and a population of about 30,000 inhabitants (Table 1). Six main ethnic groups are represented including Togeanese, Bajau, Bobongko, Buginese, Gorontalonese, and Javanese transmigrants. In some villages such as Bajau, nearly the entire working force is composed of fishers. In others, there are few full-time fishers and farming is the main activity. However, even in the latter situation most residents are at least part-time fishers. Owing to increased tourism, there is also significant employment in this economic sector with its associated support businesses. Poverty is the overwhelming economic issue throughout the islands. A total of 29 villages are classified as "poor" in which the annual income per capita is less than 700,000 Rp.

The concern for preserving the Togean Island's biologically diverse terrestrial and marine ecosystems and their importance to the economic welfare of the islands has prompted CI to develop, in partnership with YABSHI (an Indonesian non-governmental conservation organization), a long-term research and conservation program. Local project activities are administered from a permanent research station established by YABSHI at Malenge Island and are managed as a local-based entity called the Sekber Konsorsium Togean. The goal of the consortium is to develop an integrated system of marine and terrestrial reserves in which coastal communities, local government, and other stakeholders achieve consensus on matters relating to the designation, delineation, and management of the protected areas. The consortium invests considerable resources in working with local and provincial government because their programs have a huge impact on the conservation of local biodiversity. Unfortunately, the concept of environmental conservation is still poorly understood among most government agencies. YABSHI, with the support of CI, has increasingly expanded its role in community development and environmental policy decisions, at the same time building political support for conservation initiatives.

The Banggai Islands, administered by the Banggai Regency, consist of five sub-districts and 109 villages (Table 1). The approximate population of the area is 151,800. As in the Togeans, poverty is widespread and the population depends on farming and fishing for sustenance and capital. Unfortunately, there are no formal conservation programs in existence and the sustainability of natural resources is a primary concern. In both the Banggai and Togean groups the major threats to coral reefs include illegal cyanide and dynamite fishing, as well as uncontrolled clearing of forests for agriculture, and the consequent erosion and siltation of coastal waters.

Site Selection and Methods

General site locations were selected by a pre-survey consultation of marine navigation charts. The following Indonesian charts were particularly valuable in this regard: 192 (Togean Islands), 309 (Teluk Tomini), and 311 (Banggai Islands). British Admiralty chart 3240 (Teluk Tomini) was also used. An effort was made to select a representative sample of all major reef habitats, including coastal fringing reefs, sheltered lagoons, offshore barrier reefs, and atoll reefs. Exact site selection was made upon arrival at the general pre-selected area, and was influenced by weather and sea conditions.

The expedition took place aboard the *Serenade*, a 25 m, live-aboard dive boat, operated by Murex Resort, Manado, Sulawesi. The vessel was well equipped for diving, with an efficient air compressor, approximately 20 scuba tanks, and two rubber dinghies, each with a 15 hp outboard motor.

At each site, the Biological Team conducted underwater assessments that produced species lists for key coral reef indicator groups. Due to other commitments, the RAP mollusc expert, Dr. Fred Wells, was only able to participate in the first half of the survey, and mollusc data were not gathered for the Banggai Islands. General habitat information was also recorded, as was the extent of live coral cover at several depths. The main survey method consisted of direct underwater observations by diving scientists, who recorded species of corals, molluscs, and fishes. Visual transects were the main method for recording fishes and corals. Molluscs, by contrast, required the collection of live animals and shells (most released or discarded after identification). Relatively few specimens were preserved for later study, and these were invariably of species that were either too difficult to identify in the field or represented new discoveries. Further collecting details are provided in the following chapters.

Concurrently, the reef ecology and fisheries team used a 100 m line transect placed on top of the reef in three depth zones at each site to record the percentage and type of bottom cover, and abundance of selected indicator fishes for biomass estimation. In addition, observations of any obvious reef damage (bleaching, dynamite scars, etc.) were recorded. Finally, both the taxonomic and reef ecology teams recorded any observations of large vertebrates such as sharks, sea turtles, and dugong.

Reef Morphology

Four major types of reef structures are present in the Togean-Banggai area: fringing reefs, platform/patch reefs (including

cays), atolls, and barrier reefs. Fringing reefs are generally present around the periphery of islands, as well as along the peninsular mainland. Platform reefs, patch reefs, and low-lying coral cays/offshore islets are abundant and widely scattered throughout the area, including both submerged and emergent varieties. Small patch reefs are particularly common in the eastern Togeans between Talatakoh and Waleabahi islands, and off the southern coast of Peleng Island in the Banggai Group. Pasir Tengah (Site 15) is a typical "mini" atoll and the best example of this structure in the two island groups. In addition, a submerged atoll (Site 42) is present in the Banggai Islands immediately northwest of Bangkulu Island, and several partial atolls (Sites 37, 38, and 40) are located south of this island. The most extensive and best-developed barrier reef extends along the northern edge of the Togeans, about 2–3 km from shore, between Malenge Island and the western tip of Batudaka Island. There is also a less well-developed barrier reef off the southwestern corner of Peleng Island in the Banggai Group.

Major habitats are divisible into various categories depending on their degree of exposure to wind, waves and currents. Exposed seaward reefs lie at one end of the spectrum and sheltered lagoon-type reefs at the other. In between, there is a range of intermediate types that combine various proportions of outer slope and lagoon properties. The 47 surveyed sites were broadly classified as follows:

Fringing Reefs
Sheltered reefs on the leeward side of larger islands or in protected lagoons (Sites 2-4, 8, 11, 18–20, 32, 29, 31, 35, 43–44)

These sites were variable with regards to corals and other organisms depending on the degree of shelter and proximity to coastal runoff. Some areas such as Pontil Kecil Island (Site 31) were flushed by strong currents and consequently supported a wealth of corals and fishes. Others were highly sheltered in areas susceptible to siltation and consequently had poor visibility due to terrestrial runoff. In spite of their poor overall diversity, these sites often harbored luxuriant growths of hard and soft corals in shallow water.

Exposed reefs on the windward side of larger islands or the mainland peninsula (Sites 1, 6, 12–13, 17, 26–28, 30, 34, 41)

These sites typically consisted of a narrow fringing reef next to shore with a moderate to steep slope into deep water. They are generally exposed to wave action for much of the year. Coral growth is sparse in shallow water where wave-resistant forms dominate, but becomes more luxuriant with increased depth. The most spectacularly scenic reefs in this category were those at Una-una Island (Sites 12–13), which featured steep drop-offs on their outer edge.

Offshore Reefs

Platform Reefs, pinnacles, small islets/coral cays (Sites 5, 14, 25, 30, 32–33, 36, 40, 42, 47)

These sites typically arose from deep water, and either broke the surface in the form of small islands or cays, or were entirely submerged, often coming to within 2–5 m of the surface. The best example of a pristine reef, Dondola Island (Site 25), was prominent in this category. This low-lying coral cay was surrounded by luxuriant corals in shallow water, with an abrupt outer reef drop-off on its northern edge. Another richly diverse reef, characterized by a small island, surrounding reef flat, and spectacular drop-off was noted at Site 47 (Makailu Island).

Atoll reefs (Sites 15, 37–39)

These sites typically included a shallow reef crest, which gradually sloped seaward before plunging steeply to deep water. Sites in the Banggai Islands generally included a broad shallow reef flat around islands or cays before plunging steeply on the outer edge.

Barrier reefs (Sites 7, 9–10, 16, 21–22, 45–46)

These sites typically consisted of a shallow reef crest, gradual slope on the landward side, and moderate slope seaward, ending in an abrupt drop-off to deep water. Coral growth was generally luxuriant in the shallower (< 10 m) sections.

Site Locations

A total of 47 sites were visited between 24 October and 9 November 1998, and are described in detail in Technical Report 2. A summary of the sites is presented in Tables 2 and 3. Place names were taken from Indonesian marine navigation charts (92, 309, 311).

Table 2. Summary of survey sites in the Togean Islands (Sites 1–24 and mainland peninsula/shoals (Sites 25–28).

No.	Date	Location	Coordinates
1	24/10/98	S tip of Waleabahi Is.	0° 23.74' S, 122° 24.10' E
2	24/10/98	Kedodo Bay, Waleabahi Is.	0° 17.53' S, 122° 20.47' E
3	25/10/98	Lantolanto, Waleabahi Is.	0° 15.92' S, 122° 15.98' E
4	25/10/98	S tip of Waleakodi Is.	0° 17.97' S, 122° 14.07' E
5	25/10/98	Off S central Waleakodi Is.	0° 17.83' S, 122° 11.42' E
6	26/10/98	E end of Malingi Is.	0° 15.41' S, 122° 05.81' E
7	26/10/98	Malingi barrier reef	0° 12.97' S, 122° 03.89' E
8	26/10/98	Kilat Bay, N. Togean Is.	0° 19.97' S, 121° 55.44' E
9	27/10/98	N Kadidi barrier reef	0° 19.09' S, 121° 50.34' E
10	27/10/98	S Kadidi barrier reef	0° 20.35' S, 121° 49.44' E
11	27/10/98	Wakai, NE Batudaka Is.	0° 24.30' S, 121° 56.01' E
12	28/10/98	Una-una Village, Una-una Is.	0° 08.75' S, 121° 39.55' E
13	28/10/98	N tip of Una-una Is.	0° 07.49' S, 121° 38.47' E
14	28/10/98	SW corner Una-una Is.	0° 11.01' S, 121° 33.94' E
15	29/10/98	S tip of Pasir Tengah Atoll	0° 25.64' S, 121° 38.21' E
16	29/10/98	W end Batudaka barrier reef	0° 30.27' S, 121° 36.60' E
17	29/10/98	Near SW tip of Batudaka Is.	0° 34.90' S, 121° 41.68' E
18	30/10/98	Near SW tip of Batudaka Is.	0° 33.88' S, 121° 41.58' E
19	30/10/98	SW side Batudaka Is.	0° 30.97' S, 121° 44.45' E
20	30/10/98	SE side Batudaka Is.	0° 31.49' S, 121° 51.62' E
21	31/10/98	Pasir Batang barrier reef	0° 29.39' S, 122° 03.92' E
22	31/10/98	Reef near P. Mogo Besar	0° 26.12' S, 122° 00.01' E
23	31/10/98	S entrance Selat Batudaka	0° 26.60' S, 121° 56.18' E
24	1/11/98	Waleabahi - Talatakoh Is.	0° 26.37' S, 122° 16.33' E
25	1/11/98	Dondola Is.	0° 25.32' S, 122° 37.87' E
26	2/11/98	Pasirpanjang Point	0° 41.08' S, 123° 24.66' E
27	2/11/98	Puludua Is.	0° 49.36' S, 123° 27.15' E
28	2/11/98	Dongolalo Point	0° 57.35' S, 123° 28.95' E

Table 3. Summary of survey sites in the Banggai Islands.

No.	Date	Location	Coordinates
29	3/11/98	Tandah Putih, Peleng Is.	1° 13.58' S, 123° 24.29' E
30	3/11/98	Pulau Potil Kecil, Banggai Is.	1° 28.13' S, 123° 33.30' E
31	3/11/98	Banggai Harbour, Banggai Is.	1° 34.84' S, 123° 20.01' E
32	4/11/98	Bandang Is.	1° 41.12' S, 123° 27.50' E
33	4/11/98	SE end Kenau Is.	1° 46.14' S, 123° 31.36' E
34	5/11/98	N tip Kembongan Is.	1° 51.45' S, 123° 41.81' E
35	5/11/98	Lagoon W Kembongan Is.	1° 52.92' S, 123° 41.32' E
36	5/11/98	E side Tumbak Is.	1° 54.78' S, 123° 32.86' E
37	6/11/98	Atoll S of Treko Is.	2° 06.89' S, 123° 26.65' E
38	6/11/98	E side Pulau Pasibata reef	2° 07.88' S, 123° 17.40' E
39	6/11/98	SW side Saloka Is.	1° 59.42' S, 123° 15.53' E
40	7/11/98	S side Sidula Is.	1° 53.90' S, 123° 11.55' E
41	7/11/98	NE end Bangkulu Is.	1° 46.72' S, 123° 07.88' E
42	7/11/98	Atoll off NW Bangkulu Is.	1° 43.52' S, 123° 03.71' E
43	8/11/98	S Peleng near Bobo Is.	1° 38.64' S, 123° 10.27' E
44	8/11/98	Patipakaman Pt., Peleng Is.	1° 36.12' S, 123° 06.28' E
45	8/11/98	Barrier reef W Dalopo Is.	1° 36.08' S, 122° 49.62' E
46	9/11/98	SW corner Banyak Is.	1° 32.64' S, 122° 46.04' E
47	9/11/98	Makailu Is.	1° 20.73' S, 122° 49.65' E

Results and Discussion

Biological Diversity

Despite intense fishing pressure, biological diversity remains relatively high at both island groups. The number of corals, molluscs, and fishes that were recorded (Table 4) compares favorably with other parts of the "coral triangle," including areas recently surveyed by CI, such as Milne Bay, Papua New Guinea (Werner and Allen, 1998) and the Calamianes Islands, Philippines (Werner and Allen, 2000). As a result of the present RAP a total of 54 coral species were recorded from Indonesia for the first time, including at least six new species.

Although molluscs were only sampled at the Togean Islands and mainland peninsular sites, an impressive total of 541 species was recorded. Nevertheless, many more species can be expected in the area, as our inventory technique does not include the abundant micro-mollusc fauna. The area's ichthyofauna is also rich, although there was a notable lack of large reef fishes, including sharks. A total of 812 reef fishes were recorded, and an extrapolation method using six key families indicates that an overall total of at least 1,023 species can be expected. At least seven new fishes were collected during the survey. In addition, valuable observations and photographs were obtained for *Pterapogon kauderni,* a highly attractive cardinalfish endemic to the Banggai Islands that is being harvested in huge numbers for the international aquarium market. Although relatively few endemic reef organisms were recorded from the Togean and Banggai Islands, its diverse mix of local reef communities and habitats was in itself highly unique. Detailed results are given in the separate reports for corals, molluscs, and fishes.

Environmental Condition

A wide variety of environmental conditions were encountered during the survey. These ranged from nearly pristine reefs (Site 25) to areas such as site 7 that showed the effect of intense illegal fishing activity, particularly the use of explosives and cyanide. A new technique for evaluating the condition or "health" of individual sites was employed in this survey. Sites were graded according to their level of damage from both natural and human causes, in combination with overall coral cover and fish diversity. Live coral cover averaged 41–42% in the two island groups, which is relatively

Table 4. Summary of the fauna of the Togean-Banggai Islands recorded during the RAP survey.

Faunal Group	No. Families	No. Genera	No. Species
Reef corals	15	76	314
Molluscs	103	258	541
Fishes	75	272	819

high compared to many other areas of Indonesia. Dondola Island (Site 25), the reef near Lalong Village (Site 29), and the Banyak Islands (Site 46) were judged to be the best sites, being classified in the "excellent condition" category. These sites invariably had a high percentage of live coral cover, abundant fish life, and minimal signs of stress or threat.

Although the live scleractinian coral cover appears to be relatively high, there was ample evidence of reef degradation at many survey sites. The most common causes of damage were dynamite fishing and siltation due to erosion of terrestrial sediments.

Large vertebrates were generally scarce at both the Togean and Banggai Islands, undoubtedly a sign of over-exploitation. No dugong were seen, and only two sharks were noted (at Sites 19 and 45). A total of five green turtles (*Chelonia mydas*) were recorded from Sites 14, 15, 20, and 29.

Key Sites

Based on overall reef condition, biodiversity of indicator groups, and aesthetic qualities, the following sites were rated highest by RAP team members: *Togean Islands*—North Kadidi barrier reef (Sites 9–10), northeastern corner of Una-una Island (12-13), Pasir Tengah Atoll (15), western end of northern Batudaka barrier reef (16); *Peninsular reefs and shoals*—Dondola Island (25); *Banggai Islands*—Tandah Putih, Peleng Island (29), Potil Kecil Island (30), Dalopo Island barrier reef (45), Banyak Islands (46), and Makailu Island (47).

A valid case can probably be made for special protective status for the Banggai Cardinalfish, due to its extremely limited distribution and low fecundity. At the present rate of exploitation the population will decrease significantly, until its existence may be threatened. Key sites for this fish were noted at Banggai Harbor (31), Kembongan Island lagoon (35), and southern Peleng Island near Bobo Island (43). Further investigations are required to assess population numbers, but the species appears to be restricted to sheltered reefs adjacent to the larger, high islands such as Peleng and Banggai.

Community Issues

Several randomly selected villages were visited by RAP team members, who conducted informal interviews to acquire information on the relationship between marine biodiversity and the general community. An attempt was made to assess the importance of marine resources to the economic livelihood and general well-being of local villagers. Poverty seemed to be the main concern of average villagers, and particularly the effect of the continuing economic crisis, which appears to be severely impacting their livelihood. Increased prices for basic goods such as rice and medicine were of major concern. This problem has caused increased reliance on marine products for sustenance and capital, and in some cases fishers have adopted illegal methods such as dynamite

and cyanide to provide extra income to keep pace with inflation.

Conservation Issues and Recommendations

The Togean and Banggai Islands are at an important "crossroad" with regard to economic development. The monetary crisis has created a discouraging spiral of over-harvesting of marine resources in order to compensate for the increased prices of goods and services and general devaluation of the rupiah. Illegal fishing methods are common throughout the region, resulting in both depleted resources and the destruction of habitats that are vital to the sustainability of fishes and other marine organisms. There is clearly a need for better enforcement of existing fishing laws, and the initiation of sound management practices to ensure that the unique assemblage of biodiversity is preserved for future generations. From this perspective we make several recommendations:

1) **Evaluate and prevent downstream environmental impacts of land-based activities.** Fishing practices are not the only threat to the area's marine environment. Land-based threats are another concern. Uncontrolled clearing of native forests for agriculture, for example, increases erosion, which in turn introduces higher sediment loads into watercourses. These sediments are discharged into coastal waters where they can smother and kill coral reefs. The government should therefore take into account the downstream environmental impacts of any deforestation projects and subsequent land-use schemes.

   In general, watershed protection should be a primary objective of any management plan for the area. Implementing this objective will also target agricultural activities that can have negative downstream impacts on coastal and marine environments. If successfully designed and implemented, watershed protection will prevent some irreversible environmental consequences that sediment and pollution can cause in near-shore waters, while also helping rivers and streams support people and freshwater biodiversity. Watershed protection is crucial throughout the area, but especially on the large island of Peleng, which supports a wide variety of freshwater habitats.

2) **Establish marine protected areas.** The establishment of protected areas should be a primary objective of conserving the area's unique marine biological community. It is important to clarify the term "protected area" to show that it does not necessarily mean that an area or its resources is denied to a local population.

The location of protected areas should follow scientific guidelines to ensure that they cover a representative sample of the region's marine biodiversity. Target areas should include those that are exceptionally diverse, unique, and that are known to protect currently or potentially rare or endangered species, such as the Banggai Cardinalfish.

Identifying the location of marine conservation areas should be guided by an understanding of the distribution of marine biodiversity, not just locally, but also on a regional scale. Other factors also need to be considered, especially the degree of local interest and support. Whatever mechanisms are used, they should be in tune with the needs of coastal villagers.

3) **Strengthen species conservation programs for rare and endangered marine wildlife.** The Togean-Banggai area, and Indonesia in general, sustain an active trade in many species that in other parts of the world have disappeared or experienced dramatic reductions in population numbers. For example, many species of sharks have drastically declined in numbers throughout Southeast Asia and other regions where they were once common. Besides the need for protecting areas where rare, endangered, and endemic species are found, it is important to control their direct exploitation and trade. Marine species that seem particularly threatened in the Togean-Banggai area include sharks, sea turtles, dugong, giant clams, and certain fishes (e.g., Napoleon wrasse). In Indonesia, some of these animals are protected by national laws, but unfortunately these laws are rarely enforced.

At the community level, strengthening traditional values and developing more lucrative cash incentives (such as from eco-tourism projects) might provide local solutions. At the government level, there is a need for better enforcement of existing regulations and an increased capacity to conduct periodic monitoring of endangered species populations so that species conservation programs respond to changing needs. Monitoring programs should include key roles for both local people and NGOs.

4) **Enact species conservation measures for *Pterapogon kauderni*, to include an international ban on the harvest and trade of wild specimens.** In view of its highly restricted distribution, low fecundity, and current level of exploitation, the Banggai Cardinalfish should be protected by international law. It seems unlikely that local collecting bans would prove effective. The real solution for its protection would be its listing by CITES as a prohibited trade commodity. This fish is reported to breed readily in captivity and therefore the demand in the international aquarium trade could be satisfied by aquarium-bred stock. A combination of CITES listing and increased supply of captive stock would curtail the demand for wild-caught specimens.

5) **Examine feasibility of developing economic incentives that support marine conservation.** Conservation interests are frequently seen as adversarial to economic development interests. In both the Togean and Banggai Islands, conservation of marine resources will likely provide a significant economic return for its people. However, more study is needed of the socio-economic factors involved. Marine and terrestrial wildlife have major tourism potential that should not be overlooked. Worldwide, Indonesia is one of the most coveted destinations by scuba divers, but the Togean and Banggai Islands remain virtually undiscovered. In places such as Florida, reef tourism generates an estimated US $1.6 billion dollars annually (Birkland, 1997), and unlike other more destructive forms of development, can be sustainable if properly managed. Although this large volume of tourism is obviously neither possible nor desirable for the Togean-Banggai area, it does provide an example of an economic development option that concurrently respects the needs of people and conserves the environment.

Although current visitation is very low (one dive resort on Waleabahi Island, Togeans), divers should ensure that corals are not damaged by anchoring. When this activity becomes more popular it may be advisable to install mooring buoys. However, given past problems in other regions, in which buoys are removed for uses other than originally intended, installation should be carefully planned in consultation with local people and professional diving experts.

6) **Strengthen capacity within the region for managing natural environments.** It is necessary to strengthen the capacity of local government to confront the increasing environmental pressures on marine resources. Unfortunately, local governments seldom have the staff and resources to address conservation issues. Given this scenario it seems likely that national or overseas funding may be needed for any major conservation initiatives.

7) **Enforce existing laws.** Destructive fishing practices such as the use of cyanide and dynamite are illegal. However, enforcement of these bans is virtually nonexistent in areas such as the Togean and Banggai Islands. Research is needed to analyze enforcement systems available and implement the one most cost effective. Staff and equipment need to be provided to existing law

enforcement agencies for the expansion of their activities into the marine environment.

8) **Launch an environmental awareness campaign.** The results of this survey can be used to build awareness about the importance of marine conservation for the Togean and Banggai Islands. Several audiences should be targeted, especially coastal villagers, government officials, non-governmental organizations (NGOs), and international development agencies.

9) **Establish a long-term environmental monitoring program.** Periodic surveys by marine biologists are recommended to monitor the status of reef environments and of particular species, such as the Banggai Cardinalfish. As in other areas of the Indo-Pacific, these biologists might assist in the design of simple, but effective monitoring protocols which villagers, NGOs, and government officers could carry out themselves.

10) **Promote collection of data essential for marine conservation planning.** Biological data are not the only type of information that is important for conservation planning. For the Togean and Banggai Islands, layers of geophysical, political, ecological, cultural, and socio-economic information should be combined through a process that results in the definition of a regional conservation strategy. One option is to convene a workshop where a group of relevant experts and stakeholders review existing information to achieve consensus on a conservation strategy. One component of the resulting strategy would be the identification of information gaps, and proposals for how to fill them.

References

Birkland, C. (ed.). 1997. *Life and Death of Coral Reefs.* Chapman and Hall, New York.

Gosliner, T. M., D. W. Behrens, and G. C. Williams. 1996. *Coral Reef Animals of the Indo-Pacific.* Sea Challengers, Monterey, California.

Leis, J. M. 1991. Chapter 8. The pelagic stage of reef fishes: The larval biology of coral reef fishes. In: Sale, P. F. (ed.). *The Ecology of Fishes on Coral Reefs.* pp. 183–230. Academic Press, San Diego.

Lesser, M.P. 1996. Oxidative stress causes coral bleaching during exposure to elevated temperatures. *Coral Reefs* 16: 187–192.

Suharsono, R. Sukarno, M. Adrim, D. Arief, A. Budiyanto, Gyanto, A. Ibrahim, and Yahmantoro. 1995. *Weisata bahari kepulauan Banggai.* Lembaga Ilumu Pengetahuan Indonesia, Pusat Penelitian dan Pengembangan Oseanologi, Jakarta.

Wallace, C.C. and J. Wolstenholme. 1998. Revision of the coral genus *Acropora* (Scleractinia: Astrocoeniina: Acroporidae) from Indonesia. *Zoological Journal of the Linnaean Society* 123: 1–186.

Wallace, C.C. 2000. Nature and origins of the unique high diversity reef faunas in the Bay of Tomini, Central Sulawesi: The ultimate center of diversity? In: *Abstract. 9th International Coral Reef Symposium, Bali, Indonesia 2000,* D. Hopley, P. M. Hopley, J. Tamelander, T. Done (eds.). p. 46. State Ministry of the Environment, Indonesia, Indonesian Institute of Sciences, The International Society for Reef Studies.

Werner, T. B. and G.R. Allen (eds.). 1998. *A rapid biodiversity assessment of the coral reefs of Milne Bay Province, Papua New Guinea.* RAP Working Papers 11, Conservation International, Washington, DC.

Werner, T. B. and G.R. Allen, (eds.). 2001. *A rapid biodiversity assessment of the coral reefs of the Calamianes Islands, Palawan Province, Philippines.* RAP Working Papers, Conservation International, Washington, DC.

Chapter 1

Reef corals of the Togean and Banggai Islands, Sulawesi, Indonesia

Douglas Fenner

Summary

- A list of corals was compiled for 47 sites in the Togean Islands, Banggai Islands, and penisular reefs and shoals separating the two areas. These included 24 sites in the Togean Islands, 4 mainland penisule/shoals, 18 in the Banggai Islands, and one on a shoal that separates the two island groups. The survey involved 51 hours of scuba diving by D. Fenner to a maximum of 44 meters.

- The Togean and Banggai Islands have a diverse coral fauna. A total of 314+ species were observed or collected during the survey.

- Species numbers at visually sampled sites ranged from 34 to 106, with an average of 70 per site. Togean sites had slightly more species (average 74 per site) than the Banggai sites (average 67 per site), but the difference was not significant.

- Despite a greater sampling effort (28 versus 19 sites) in the Togeans, slightly more species were recorded from the Banggai Islands (262 and 267 species, respectively).

- *Acropora, Montipora,* and *Porites* were dominant, with 49, 21, and 12 species respectively; this is typical of Indo-Pacific reefs.

- The overwhelming majority (94%) was zooxanthellate Scleractinia, with only a few non-sclearctinian and azooxanthellate species, as is typical of Indo-Pacific reefs.

- At least six new species were collected during the survey: one species each of *Acropora, Porites, Leptoseris, Echinophyllia,* and two of *Galaxea.* In addition, 55 species were recorded from Indonesia for the first time.

Introduction

The principle aim of the survey was to provide an inventory of the coral species growing on reefs and associated habitats, including those growing on sand or other soft sediments within and around reefs. The primary group of corals is the zooxanthellate scleractinian corals, those containing single-cell algae and which contribute to reef building. Also included are a few zooxanthellate non-scleractinian (e.g., blue coral, organ-pipe coral, and fire coral), azooxanthellate scleractinian corals (*Tubastrea* and a few others), and azooxanthellate non-scleractinian corals (*Distichopora* and *Stylaster*). All produce calcium carbonate skeletons, which contribute to reef building to some degree.

The results of this survey facilitate a comparison of the faunal richness of the Togean and Banggai Islands with other parts of Southeast Asia and adjoining regions. However, the list of corals presented below is still incomplete, due to the time restriction of the survey (16.5 days), the patchy distribution of many corals, and the difficulty in identifying some species underwater. Corals are sufficiently difficult to identify that there are often significant differences between the opinions of leading experts.

Methods

The author surveyed corals during 51 hours of scuba, diving to a maximum depth of 44 m. A list of coral species was compiled at 47 sites. The basic method consisted of underwater observations, usually during a single, 60–90 minute dive at each site. The name of each species identified was marked on a plastic sheet on which species names were printed. A direct descent was made in most cases to the base of the reef, to or beyond the deepest coral visible. Most of the dive consisted of a slow ascent along the reef in a zigzag path to the shallowest point. All habitats encountered were surveyed, including sandy areas, walls, overhangs, slopes, and shallow reef. Areas typically hosting few or no corals, such as seagrass beds and mangroves, were not surveyed.

Many corals can be positively identified underwater to species level, but several are impossible to confirm without seeing the living polyps. Field guides assisted identification (Veron, 1986; Nishihira, 1991; and Nishihira and Veron, 1995); however, some corals were dependent on laboratory examination for positive identification. This was aided by the use of various references including Best and Suharsono (1991), Cairns and Zibrowius (1997), Claereboudt (1990), Dai (1989), Dai and Lin (1992), Dineson (1980), Hodgson (1985), Hodgson and Ross (1981), Hoeksema (1989), Hoeksema and Best (1991 and 1992), Moll and Best (1984), Nemenzo (1986), Ogawa and Takamashi (1993 and 1995), Sheppard and Sheppard (1991), Veron (1985, 1990, and in press), Veron and Pichon (1976, 1980, and 1982), Veron et al. (1977), Wallace (1994 and 1997a), and Wallace and Wolstenholme (1998).

Results

A total of 314+ species in 75 genera of stony corals (295+ species and 66 genera of zooxanthellate Scleractinia) were recorded, with 262 species in the Togeans and 267 in the Banggai Group (Appendix 1). Many of these species are illustrated in Veron (1986) or Nishihira and Veron (1995), and nearly all are illustrated in Veron (2000a), and Veron and Stafford-Smith (in press).

General faunal composition

The coral fauna consists mainly of Scleractinia. The genera with the largest numbers of species found were *Acropora*,

Table 1.1 Genera with the greatest number of species.

Rank	Genus	Species
1	*Acropora*	49
2	*Montipora*	21
3	*Porites*	12
4	*Fungia*	11
5	*Pavona*	10
6	*Leptoseris*	8
7	*Lobophyllia*	7
8	*Echinopora*	7
9	*Favia*	7
10	*Pectinia*	6

Montipora, Porites, Fungia, Pavona, Leptoseris, Lobophyllia, Echinopora, Favia, and *Pectinia*. These 10 genera account for about 44% of the total observed species (Table 1.1).

The order of the most common genera is typical of Western Pacific reefs (Table 1.2), with a few minor differences; *Acropora, Montipora*, and *Porites* are always the three most speciose genera. The farther down the list one moves, the more variable the order becomes, with both the number of species and the differences between genera decreasing.

Most of the corals were zooxanthellate (algae-containing, reef-building) scleractinian corals, with 94% of the species in this group. Seven corals were azooxanthellate (lacking algae) scleractinians, making up 2% of the total, and the remaining 12 species (4%) were non-scleractinians.

Table 1.2. Genera with the greatest number of species for various sites in the West Pacific: eastern and western Australia (E. Aus. and W. Aus.), Philippines (Phil.), Japan (all from Veron, 1993), Calamianes Islands (Calam.), Philippines (Werner and Allen, 2001) and North Sulawesi (N. Sul.) (present survey).

Rank	Genus	% of Fauna					
		E. Aus	W. Aus.	Phil.	Japan	Calam.	N. Sul.
1	*Acropora*	19	18	17	19	13	16
2	*Montipora*	9	10	10	9	7	5
3	*Porites*	5	4	6	6	3	3
4	*Favia*	4	4	4	4	3	2
5	*Goniopora*	4	4	3	4	1	1
6	*Fungia*	4	3	4	3	4	3
7	*Pavona*	2	3	3	3	4	2
8	*Leptoseris*	2	2	2	3	4	2
9	*Cycloseris*	3	2	3	2	1	1
10	*Psammocora*	2	3	2	2	2	2

Zoogeographic affinities of the coral fauna

The reef corals of the Togean and Banggai Islands, and Sulawesi in general, belong to the overall Indo-West Pacific faunal province. A few species span the entire range of the province, but most do not. The Togean and Banggai Islands lie within the central area of greatest marine biodiversity, referred to as the Coral Triangle, which encompasses central Indonesia and the Philippines. Coral diversity decreases in all directions from the Coral Triangle. For example, 80 species occur at an island near Tokyo, 65 species at Lord Howe Island, off southeastern Australia, about 45 species at the Hawaiian Islands, and about 20 species on the Pacific coast of Panama. Species attenuation is significantly less to the west in the Indian Ocean and Red Seas, though this area, like many others, is insufficiently studied to provide accurate figures.

Corals are habitat-builders and appear to have less niche-specialization than some other groups, occurring over a relatively wide range of exposure and light intensity. Thus, there are a few corals that are restricted to zones such as very shallow areas, protected areas, deep water, shaded niches, soft bottoms, or exposed areas. However, limited zonation occurs that is correlated with depth and wave/current exposure. Corals are primarily autotrophic, relying on the photosynthetic products of their symbiotic algae, supplemented by plankton caught by either filter or suspension feeding. Most require hard substrate for attachment, but a few thrive on soft substrates.

During the current survey barrier reefs supported the greatest number of coral species, followed in order by platform reefs, atolls, exposed fringing reefs, and sheltered fringing reefs (Table 1.3).

Most corals found in this area have fairly wide distributions within the Indo-Pacific. A majority of species have a pelagic larval stage, with a minimum of a few days pelagic development for broadcast spawners (most species), and larval settling competency lasting for at least a few weeks. A minority of corals release brooded larvae that, depending on the species, may settle immediately or exhibit pelagic dispersal stages of variable length.

The main zoogeographic categories for Togean-Banggai corals have relatively broad distributions that extend beyond Indonesia: 78% have ranges that extend both west and east of Indonesia, 10% extend east but not west from Indonesia,

Table 1.3. Species richness (average number of species per site) according to type of reef.

Reef Type	No. Species
Barrier	90
Platform	84
Atolls	74
Exposed Fringing	71
Sheltered Fringing	69

about 5% range west, north, or south, and 7% are restricted to the Coral Triangle. None are endemic to Indonesia (based on Veron, 1993). Wallace (1997a) listed 10 species of *Acropora* as Indonesian endemics. However, subsequent collections from surrounding regions have reduced the number to only two species, or 2.4% of the total Indonesian *Acropora* fauna.

Comparison between coral faunas of the Togean and Banggai Islands

Of the 314+ species, 45 were found only in the Togeans and 50 only in the Banggais, with 219 found in both. The complete list by sites is presented in Appendix 1. Sites 46, 24, and 40 had the highest species richness, with 100, 93, and 92 species respectively. Sites 35, 11, 10, and 18 had the lowest species richness, with 28, 44, 50, and 51 species. The 10 richest sites are summarized in Table 1.4, and the number of species at all sites is presented in Table 1.5.

The majority of species were recorded at the first 15 sites. Subsequent species were added to the list at a slow, but relatively steady rate, indicating that sufficient sites were surveyed (Figure 1.1).

Togean sites had a slightly higher average number of species per site than Banggai sites, 74 versus 67, but the difference was not significant (two-tailed t-test with unequal variances, $p = 0.11$). This result is consistent with the lack of a difference in the total number of corals (262 in the Togeans and 267 in the Banggai Group).

A few corals showed striking differences in abundance between the two areas. The wide-ranging *Scolymia vitiensis* was found in low numbers at 16 sites in the Togeans, but was absent in the Banggai Group. Similarly, *Acropora togianensis,* which was described from the Togeans, was found on 11 sites there, and was common and conspicuous on many. It was not seen at any sites in the Banggai Islands despite intensive collecting efforts in Indonesia and surrounding regions by Wallace and Veron (pers. comm.).

One of the rarest species found was *Acropora russelli,* previously known from Cartier Reef, off northwestern Australia, and Halmahera, Indonesia. Likewise, *Leptoseris amitoriensis,* previously known from one site in Japan and another at the Calamianes Islands, Philippines, was another extremely rare species collected during the present survey

Wallace (1997b) reported 53 species of *Acropora* from the Togean Islands, compared to 28–61 species (mean = 50.4 species) in four other areas of Indonesia. In the present survey, 42 species of *Acropora* were found in the Togeans, and 51 species in the combined Togean-Banggai area. The latter figure is perhaps more comparable with Wallace's total con-

Table 1.4. Sites with the highest coral diversity.

Rank	Site	No. Species
1	46	100
2	24	93
3	40	92
4	45	89
5	42	89
6	9	87
7	38	86
8	41	86
9	21	85
10	16	84

Table 1.5. Number of coral species observed at each site.

Site	Species	Site	Species	Site	Species
1	64	17	51	33	69
2	52	18	51	34	56
3	69	19	73	35	29
4	52	20	83	36	67
5	64	21	84	37	64
6	57	22	80	38	86
7	68	23	57	39	60
8	54	24	93	40	92
9	87	25	76	41	87
10	50	26	75	42	89
11	44	27	75	43	73
12	69	28	75	44	87
13	67	29	61	45	89
14	70	30	69	46	100
15	84	31	70	47	83
16	62	32	71		

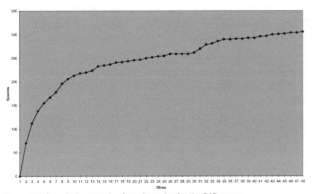

Figure 1.1. Cumulative records of coral species for the RAP survey.

sidering that two persons were involved with her collecting effort and only *Acropora* species were investigated. Moreover, Wallace collected extensive samples in contrast to the present survey, which was primarily visual. Wallace (1999a) further increased the Togeans total to 61 species, stating that this island group was one of the most diverse areas known for this genus.

Species of special interest
Laboratory examination of the approximately 50 coral species collected during this survey revealed at least six new species (Table 1.6) and further study of the material may yield additional undescribed taxa.

A number of new species were described during this survey. Type specimens were collected, and all were described in Veron (2000a) with photos taken by Fenner during this survey.

Acropora cylindrica Veron and Fenner, 2000 forms branching staghorn colonies with smooth cylindrical branches that taper smoothly to a rounded tip. Radial corallites are just slightly exsert, and their white color contrasts with the brown branch surface. Appears similar to *Acropora togianensis* but without tubercles between corallites. Also known from Papua New Guinea.

Porites rugosa Fenner and Veron, 2000a forms bushy branching colonies with very rough branches, purple or orange with yellow branch tips. Found at Sites 17, 19, and 21.

Leptoseris striata Fenner and Veron, 2000b forms small circular dishes, with a large central corallite and small streaks on its surface. Found at Sites 2, 15 and 47. Also known from the Philippines.

Galaxea longisepta Fenner and Veron, 2000c forms small encrusting patches, only about 10 cm diameter, at medium depths, often on a steep slope or partly shaded location. Corallites are widely spaced, tall, and have extremely exsert septa. Found at Sites 1, 4, 7, 12–15, 20, 21, 24, 25, and 35. The type specimen was collected on this expedition. Also known from the Philippines and Great Barrier Reef.

Galaxea cryptoramosa Fenner and Veron, 2000d forms lumps about 20–30 cm diameter in relatively shallow water. The corallites appear much as in *Galaxea fasicularis,* but colonies can be broken apart to reveal that they are composed of irregular branches with corallites along their length. Found at Sites 30, 32, and 34. The type specimen was collected on this expedition.

Echinophyllia costata Fenner and Veron, 2000e forms large plates growing upward at about a 45° angle in concentric circles like rose petals. Corallites on the thin plates look like those of *Favia* on living colonies, but examination of the skeleton reveals that they are *Echinophyllia*. It lives at medium depths in fields of other foliose species. Found at Sites 35 and 43.

Table 1.6. New species observed during survey, but found first elsewhere. Species marked with * denote need for further study to be confirmed as new taxa.

Species	Region
Seriatopora dendritica[1] Veron, 2000b	Coral Triangle, Papua New Guinea
Montipora delicatula Veron, 2000c	Coral Triangle
Montipora sp.1	
Montipora palawanensis Veron, 2000d	Coral Triangle
Montipora verruculosus Veron, 2000e	Coral Triangle
Acropora filiformis Veron, 2000f	Calamianes, Philippines
Halomitra meierae Veron and Maragos, 2000	Bali, Indonesia
Lobophyllia flabelliformis Veron, 2000g	Coral Triangle, Ryukyu Islands, Papua New Guinea, Northeast Australia
Favia truncatus Veron, 2000h	Southeast Asia, Northern and Northeastern Australia, Indian Ocean, Madagascar
Platygyra acuta Veron, 2000i	Coral Triangle, Indian Ocean, Madagascar, Red Sea
Montastrea colemani Veron, 2000j	Coral Triangle, Northeastern Australia, Madagascar, Gulf of Aden, Melanesia
**Dendrophyllia* sp. 1	
**Heliopora* sp. 1	
**Tubipora* sp. 1	
**Tubipora* sp. 2	

[1] The species name is given as both "dentritica" and "dendritica" in Veron (2000a). We presume that the former is a typographical error and the latter the correct name

In addition, 15 species of corals that were still undescribed at the time of the survey were recorded. However, most of these were subsequently described by Veron (2000a), who discovered them elsewhere (Table 1.6). *Montipora* sp. 1, as well as the last four entries in the table were collected by Fenner in the Philippines and may be described at a later date if confirmed as new taxa.

Overview of the Indonesian coral fauna

The Indonesian coral fauna is no doubt one of the richest in the world. The only other country with comparable diversity is the Philippines, although the New Guinea region (comprised of Papua New Guinea and Papua Province of Indonesia) may have comparable richness.

The total number of species found in this study, 314+, is slightly less than the number (334) reported in the most thorough study of Indonesian corals to date (Best et al., 1989). Previous Indonesian coral studies include those on reefs in the Jakarta area by Moll and Suharsono (1986) and Brown et al. (1985), which recorded 193 and 88 species respectively. In addition, Jonker and Johan (1999) recorded 163 species from Sumatra based on a study of museum specimens.

A total of 40 previously described species were recorded from Indonesian waters for the first time during the present survey (Table 1.7). Eight of these (*Acropora fastigata, Porites cumulatus, Montipora confusa, Montipora florida, Montastrea salebrosa, Pachyseris foliosa, Oxypora crassispinosa,* and *Euphyllia paradivisa*) were formerly known only from the

Philippines, and *Leptoseris amitoriensis* was known only from Japan. The latter species as well as *Pachyseris involuta* and *Symphallia hassi* are extremely rare in collections.

Table 1.7. New records of corals not previously reported from Indonesia.

Species	Species
Pocillopora ankeli	*Pachyseris foliosa*
Montipora cactus	*Pachyseris gemmae*
Montipora capitata	*Pachyseris involuta*
Montipora confusa	*Fungia klunzingeri*
Montipora florida	*Galaxea paucisepta*
Montipora hirsuta	*Pectinia teres*
Montipora samarensis	*Mycedium mancaoi*
Acropora fastigata	*Oxypora crassispinosa*
Astreopora randalli	*Oxypora glabra*
Astreopora suggesta	*Lobophyllia robusta*
Porites attenuata	*Symphyllia hassi*
Porites cumulatus	*Leptoria irregularis*
Porites evermanni	*Montastrea salebrosa*
Porites monticulosa	*Euphyllia paraancora*
Porites negrosensis	*Euphyllia paradivisa*
Porites sillimaniana	*Euphyllia yaeyamensis*
Goniopora pendulus	*Turbinaria irregularis*
Pavona bipartita	*Stylaster* sp. 1
Pavona minuta (=xarife)	*Stylaster* sp. 2
Leptoseris amitoriensis	*Distichopora nitida*

The combination of 15 undescribed species first collected outside of Indonesia, 40 described species representing new records for the country, and at least six new species, gives an overall total of 61 new additions to the Indonesian fauna. Including the present report, 478 species of Scleractinia (and a total of 522 species of stony corals) have been reported from Indonesia (Table 1.8). Additional species are undoubtedly in unpublished records held by J. E. N. Veron. Thus, central Indonesia, including Sulawesi, is clearly within the area of maximum coral diversity in the western Pacific. Corals have been investigated most extensively in the Philippines. The most recent published count (Veron and Hodgson, 1989) of 411 Scleractinia is similar to that for Indonesia, but additional unpublished data has significantly increased the total. However, further study is required in both countries to obtain comprehensive lists, and there are no complete species listings for any local areas in the region.

The summaries provided in Table 1.8 are based on relatively recent literature. The review by Tomascik et al. (1997) was particularly useful, as it includes summary data from previous studies such as the comprehensive review of Best et al. (1989). Earlier studies were not considered, largely because of the many difficulties associated with misidentifications and synonyms.

It is interesting to compare the present data with that of the previous RAP survey in the Calamianes Islands (Palawan Province, Philippines) (Werner and Allen, 2001). The total number of species was similar: 311 species in Sulawesi and 304 species in the Calamianes. Likewise, 76 genera were found in Indonesia and 75 genera in the Calamianes. However, an average of 92.5 species were found at each site in the Calamianes, compared with 69.7 species per site in Sulawesi. This difference is highly significant ($p < 0.00001$, two-tail t-test with unequal variances; 2 sites with 2 dives each in the Philippines excluded). Further, variance in the number of species per site was much higher for the Philippines, with the standard deviation 19.7, as opposed to 9.4 in Sulawesi. The latter is easily explained by a more standard search procedure in Sulawesi compared to the Calamianes, where some dives were largely spent collecting new species. Although Sulawesi had roughly the same total number of species, there were fewer species per site compared with the Calamianes. There were 47 sites at Sulawesi compared to 37 at the Calamianes. A total of 304 species were recorded in the Calamianes after 37 sites compared to 291 after 37 sites in Sulawesi. Sulawesi had 13 fewer species, or a 4% difference. Sulawesi had 23 fewer species per site, 25% of the number of species per site in the Calamianes. Together, these data indicate that the sites in the Philippines were slightly richer than those in Sulawesi.

References

Best, M.B. and B.W. Hoeksema. 1987. New observations on Scleractinian corals from Indonesia: 1. Free-living species belonging to the Faviina. *Zoologishe Mededelingen Leiden.* 61: 387–403.

Best, M.B., B.W. Hoeksema, W. Moka, H. Moll, Suharsono and I. Nyoman Sutarna. 1989. Recent scleractinian coral species collected during the Snellius-II Expedition in eastern Indonesia. *Netherlands Journal of Sea Research* 23: 107–115.

Best, M.B. and Suharsono. 1991. New observations on Scleractinian corals from Indonesia: 3. Species belonging to the Merulinidae with new records of *Merulina* and *Boninastrea. Zoologishe Mededelingen Leiden* 65: 333-342.

Boschma, H. 1959. Revision of the Indo-Pacific species of the genus *Distichopora. Bijdragen tot de Dierkunde* 29: 121–171.

Brown, B.E., M.C. Holley, L. Sya'rani and M. Le Tissier. 1985. Coral diversity and cover on reef flats surrounding Pari Island, Java Sea. *Atoll Research Bulletin* 281: 1–17.

Cairns, S.D. and H. Zibrowius. 1997. Cnidaria Anthozoa: Azooxanthellate Scleractinia from the Philippines and Indonesian regions. In: A. Crozier and P. Bouchet (eds.), *Resultats des Campagnes Musorstom*, Vol 16, Mem. Mus. Nat. Hist. 172: 27–243.

Table 1.8. Numbers of coral species reported from Indonesian shallow waters (stony corals includes zooxanthellate scleractinia, shallow water azooxanthellate scleractinia, stony octocorals and hydrocorals).

| | | Cumulative Total Species | | |
	Total Spp.	New Records	Stony Corals	Zooxan. Scleract.
Tomascik et al, 1977	405	405	405	389
Wallace 1994, 1997a, 1998, 1999b	94	31	436	420
Cairns & Zibrowius, 1997	34	25	461	—
This report	314+	61+	522+	478+

Claereboudt, M. 1990. *Galaxea paucisepta* nom. nov. (for *G. pauciradiata*), rediscovery and redescription of a poorly known scleractinian species (Oculinidae). *Galaxea* 9: 1–8.

Dai, C-F. 1989. Scleractinia of Taiwan. I. Families Astrocoeniidae and Pocilloporiidae. *Acta Oceanographica Taiwanica* 22: 83–101.

Dai, C-F. and C-H. Lin. 1992. Scleractinia of Taiwan III. Family Agariciidae. *Acta Oceanographica Taiwanica* 28: 80–101.

Dineson, Z.D. 1980. A revision of the coral genus *Leptoseris* (Scleractinia: Fungiina: Agariciidae). *Memoires of the Queensland Museum* 20: 181–235.

Fenner, D. and J.E.N. Veron. 2000a. Family Poritidae, genus *Porites, rugosa*. In: *Corals of the World Vol. 3*. J.E.N. Veron, p. 342. Australian Institute of Marine Science, Townsville.

Fenner, D. and J.E.N. Veron. 2000b. Family Agariciidae, genus *Leptoseris, striata*. In: *Corals of the World Vol. 2*. J.E.N. Veron, p. 212. Australian Institute of Marine Science, Townsville.

Fenner, D. and J.E.N. Veron. 2000c. Family Oculinidae, genus *Galaxea, longisepta*. In: *Corals of the World Vol. 2*. J.E.N. Veron, p. 116–117. Australian Institute of Marine Science, Townsville.

Fenner, D. and J.E.N. Veron. 2000d. Family Oculinidae, genus *Galaxea, cryptoramosa*. In: *Corals of the World Vol. 2*. J.E.N. Veron, p. 114. Australian Institute of Marine Science, Townsville.

Fenner, D. and J.E.N. Veron. 2000e. Family Pectiniidae, genus *Echinophyllia, costata*. In: *Corals of the World Vol. 2*. J.E.N. Veron, p. 330. Australian Institute of Marine Science, Townsville.

Hodgson, G. 1985. A new species of *Montastrea* (Cnidaria, Scleractinia) from the Philippines. *Pacific Science* 39: 283–290.

Hodgson, G. and M.A. Ross. 1981. Unreported scleractinian corals from the Philippines. *Proceedings of the Fourth International Coral Reef Symposium*, 2: 171–175.

Hoeksema, B.W. 1989. Taxonomy, phylogeny and biogeography of mushroom corals (Scleractinia: Fungiidae). *Zoologishe Verhandelingen* 254: 1–295.

Hoeksema, B.W. 1992. The position of northern New Guinea in the center of marine benthic diversity: a reef coral perspective. *7th International Coral Reef Symposium* 2: 710–717.

Hoeksema, B.W. and M.B. Best. 1991. New observations on scleractinian corals from Indonesia: 2. Sipunculan-associated species belonging to the genera *Heterocyathus* and *Heteropsammia*. *Zoologishe Mededelingen* 65: 221–245.

Jonker, L. and O. Johan. 1999. Checklist of the scleractinian coral species from the waters of Padang (West Sumatra, Indonesia) held in the coral collection of Bung Hatta University. *The Beagle, Records of the Museums and Art Galleries of the Northern Territory* 15: 47–54.

Moll, H. and M.B. Best. 1984. New scleractinian corals (Anthozoa: Scleractinia) from the Spermonde Archipelago, south Sulawesi, Indonesia. *Zoologische Mededelingen* 58: 47–58.

Moll, H. and Suharsono. 1986. Distribution, diversity and abundance of reef corals in Jakarta Bay and Kepulauan Seribu. *UNESCO Reports in Marine Science* 40: 112–125.

Nemenzo, F. Sr. 1986. *Guide to Philippine Flora and Fauna: Corals*. Natural Resources Management Center and the University of the Philippines. 273 pp.

Nishihira, M. 1991. *Field Guide to Hermatypic Corals of Japan*. Tokai University Press, Tokyo. 264 pp. (in Japanese)

Nishihira, M. and J. E. N Veron. 1995. *Corals of Japan*. Kaiyusha Publishers Co., Ltd, Tokyo. 439 pp. (In Japanese)

Ogawa, K., and K. Takamashi. 1993. A revision of Japanese ahermatypic corals around the coastal region with guide to identification- I. Genus *Tubastraea*. Nankiseibutu: *The Nanki Biological Society* 35: 95–109. (in Japanese)

Ogawa, K. and K. Takamashi. 1995. A revision of Japanese ahermatypic corals around the coastal region with guide to identification- II. Genus *Dendrophyllia*. Nankiseibutu: *The Nanki Biological Society* 37: 15–33. (in Japanese)

Sheppard, C.R.C. and A.L.S. Sheppard. 1991. Corals and coral communities of Arabia. *Fauna of Saudi Arabia* 12: 3–170.

Tomajcik, T., A. J. Mah, A. Montji, and N. K. Moosa 1997 The Ecology of Indonesia: Seas.

Veron, J.E.N. 1985. New scleractinia from Australian reefs. *Records of the Western Australian Museum* 12: 147–183.

Veron, J.E.N. 1986. *Corals of Australia and the Indo-Pacific.* University of Hawaii Press. 644 pp.

Veron, J.E.N. 1990. New scleractinia from Japan and other Indo-West Pacific countries. *Galaxea* 9: 95–173.

Veron, J.E.N. 1993. A biogeographic database of hermatypic corals. *Australia Institute of Marine Science Monograph* 10: 1–433.

Veron, J.E.N. 2000a. *Corals of the World.* Australian Institute of Marine Science, Townsville.

Veron, J.E.N. 2000b. Family Pocilloporidae, genus *Seriatopora, dendritica.* In: *Corals of the World Vol. 2.* J.E.N. Veron, p. 46–47. Australian Institute of Marine Science, Townsville.

Veron, J.E.N. 2000c. Family Acroporidae, genus *Montipora, delicatula.* In: *Corals of the World Vol. 1.* J.E.N. Veron, p. 70–71. Australian Institute of Marine Science, Townsville.

Veron, J.E.N. 2000d. Family Acroporidae, genus *Montipora, palawanensis.* In: *Corals of the World Vol. 1.* J.E.N. Veron, p. 132. Australian Institute of Marine Science, Townsville.

Veron, J.E.N. 2000e. Family Acroporidae, genus *Montipora, verruculosus.* In: *Corals of the World Vol. 1.* J.E.N. Veron, p. 136. Australian Institute of Marine Science, Townsville.

Veron, J.E.N. 2000f. Acroporidae, genus *Acropora, filiformis.* In: *Corals of the World Vol. 1.* J.E.N. Veron, p. 418. Australian Institute of Marine Science, Townsville.

Veron, J.E.N. 2000g. Family Mussidae, genus *Lobophyllia, flabelliformis.* In: *Corals of the World Vol. 3.* J.E.N. Veron, p. 48–49. Australian Institute of Marine Science, Townsville.

Veron, J.E.N. 2000h. Family Faviidae, genus *Favia, truncatus.* In: *Corals of the World Vol. 3.* J.E.N. Veron, p. 113. Australian Institute of Marine Science, Townsville.

Veron, J.E.N. 2000i. Family Faviidae, genus *Platygyra, acuta.* In: *Corals of the World Vol. 3.* J.E.N. Veron, p. 190. Australian Institute of Marine Science, Townsville.

Veron, J.E.N. 2000j. Family Faviidae, genus Favidae, *Montastrea, colemani.* In: *Corals of the World Vol. 3.* J.E.N.

Veron, p. 219. Australian Institute of Marine Science, Townsville.

Veron, J.E.N. and D. Fenner. 2000. Family Acroporidae, genus *Acropora, cylindrica.* In: *Corals of the World Vol. 1.* J.E.N. Veron, p. 193. Australian Institute of Marine Science, Townsville.

Veron, J.E.N. and G. Hodgson. 1989. Annotated checklist of the hermatypic corals of the Phillipines. *Pacific Science* 43: 234–287.

Veron, J.E.N. and J. E. Maragos. 2000. Family Fungiidae, genus *Halomitra, meierae.* In: *Corals of the World Vol. 2.* J.E.N. Veron, p. 300. Australian Institute of Marine Science, Townsville.

Veron, J.E.N. and M. Pichon. 1976. Scleractinia of Eastern Australia. I. Families Thamnasteriidae, Astrocoeniidae, Pocilloporidae. *Australian Institute of Marine Science Monograph Series* 1: 1–86.

Veron, J.E.N. and M. Pichon. 1980. Scleractinia of Eastern Australia. III. Families Agariciidae, Siderastreidae, Fungiidae, Oculilnidae, Merulinidae, Mussidae, Pectiniidae, Caryophylliidae, Dendrophylliidae. *Australian Institute of Marine Science Monograph Series* 4: 1–422.

Veron, J.E.N. and M. Pichon. 1982. Scleractinia of Eastern Australia. IV. Family Poritidae. *Australian Institute of Marine Science Monograph Series* 5: 1–210.

Veron, J. E. N., M. Pichon, and M. Wijsman-Best. 1977. Scleractinia of Eastern Australia. II. Families Faviidae, Trachyphylliidae. *Australian Institute of Marine Science Monograph Series* 3: 1–233.

Veron, J.E.N. and C. Wallace. 1984. Scleractinia of Eastern Australia. V. Family Acroporidae. *Australian Institute of Marine Science Monograph Series* 6: 1–485.

Wallace, C.C. 1994. New species and a new species-group of the coral genus *Acropora* (Scleractinia: Astrocoeniina: Acroporidae) from Indo-Pacific locations. *Invertebrate Taxonomy* 8: 961–88.

Wallace, C.C. 1997a. New species of the coral genus *Acropora* and new records of recently described species from Indonesia. *Zoological Journal of the Linnean Society* 120: 27–50.

Wallace, C.C. 1997b. The Indo-Pacific centre of coral diversity re-examined at species level. *Proceedings of the 8th International Coral Reef Symposium* 1: 365–370.

Wallace, C.C. 1999a. The Togian Islands: Coral reefs with a unique coral fauna and an hypothesized Tethys Sea signature. *Coral Reefs* 18: 162.

Wallace, C.C. 1999b. *Staghorn Corals of the World, a Revision of the Genus Acropora*. CSIRO Publishers, Collingwood, Australia. 422 pp.

Wallace, C.C. and J. Wolstenholme. 1998. Revision of the coral genus *Acropora* in Indonesia. *Zoological Journal of the Linnean Society* 123: 199–384.

Chapter 2

Condition of Coral Reefs in the Togean and Banggai Islands, Sulawesi, Indonesia

Syafyudin Yusuf and Gerald R. Allen

Summary

- Reef condition is a term pertaining to the general "health" of a particular site as determined by assessment of key variables including natural and human-induced environmental damage, and general biodiversity as defined by major indicator groups (corals and fishes).

- Reef condition was assessed at 47 sites in the Togean Islands, Banggai Islands, and peninsular reefs and shoals separating the two areas. As part of this analysis, substrate cover was recorded in three depth zones (2–6 m, 8–14 m, 18–26 m) at each site along a 100 m transect.

- A Reef Condition Index (RCI) value was calculated for each site. Essentially it is derived from three equally weighted components: coral diversity, fish diversity, and relative damage from human and natural causes. The latter component also incorporates the percentage of live coral cover.

- The hypothetical maximum RCI for a pristine reef is 300; RCI values are useful for interpreting reef condition and comparing sites. Depending on their RCI, sites can be classified as extraordinary, excellent, good, moderate, poor, and very poor. The frequency of Togean-Banggai sites was as follows: extraordinary (0), excellent (2), good (10), moderate (19), poor (14), and very poor (2).

- Dondola Island (Site 25), Tandah Putih, Peleng Island (Site 29), and the Banyak Islands (Site 46) were judged to be the best sites in terms of reef condition. These sites had a high percentage of live coral cover, abundant fish life, and showed minimal signs of stress or threat.

- The percentage of live coral cover for survey sites averaged 41–42%. A total of 13 sites had an average in excess of 50%. These values compare favorably with other surveyed areas in Indonesia.

- Coral bleaching was observed at nearly every site in the Togean Islands, but at only one site in the Banggai Islands. Widespread bleaching in the Togeans appears to be correlated with increased sea temperatures which ranged from 30–33°C.

- Crown-of-thorns starfish *(Acanthaster planci)*, a well-documented predator of scleractinian corals, were observed at nine sites in the Togean Islands and five sites in the Banggai Islands. They were generally seen in low numbers (1–2 per dive) except at Sites 6 and 43, where up to 30 individuals were observed.

- Illegal fishing practices, including the use of dynamite, cyanide, and various types of netting, are common in both island groups.

- Dynamite damage was observed at 86% of sites in the Togean and Banggai Islands. In most cases the damage was light to moderate, but severe destruction was noted at sites 41 and 42 in the Banggai Islands.

Introduction

Coral reefs are an integral part of the marine community in Indonesian seas. The archipelago contains more than 15,000 islands, of which nearly all support at least some populations of scleractinian corals (Soekarno et al., 1983). Indeed coral reefs provide shelter and feeding opportunities for a wealth of diverse organisms, and are therefore considered one of the most important marine ecosystems. The condition of live corals at a particular site or region is indicative of overall reef "health" and it is therefore desirable to assess this feature for conservation purposes. Hence, a vital part of marine RAP surveys involves an assessment of live coral cover and various threats, past or present, which may exert deleterious effects on this fragile ecosystem.

Reefs throughout the "Coral Triangle," and Indonesian reefs in particular, are being depleted of their biotic resources at an alarming rate (Nontji, 1986). The widespread use of dynamite and potassium cyanide is a major problem throughout the region and no area appears to be exempt. Surprisingly, some of the worst affected areas lie within marine national parks and other so-called protected areas.

Detailed investigations of Indonesian reefs have been few and far between with a few notable exceptions including work in the eastern part of the Archipelago by the Snellius II Expedition of 1984–1985 (Suharsono et al. 1985 and Soekarno 1989). In addition, a comprehensive reef-study program has been initiated at Takabonerate National Marine Park in South Sulawesi and at the Banggai Islands off Central Sulawesi (Adhyakso, 1997). Finally, plans for extensive reef investigations are currently being developed by the COREMAP program of LIPI (LIPI, 1998; COREMAP, 1998).

The primary goal of the present marine RAP survey was to contribute to overall knowledge of Indonesian coral reefs and associated biodiversity and therefore provide useful information to assist conservation planning and management.

Methods

Definition of Reef Condition

Reef condition is used here as a term that reflects the general "health" of a particular site as determined by an assessment of variables that include environmental damage due to natural and human causes, and general biodiversity as defined by key indicator groups (corals and fishes). It also takes into account amounts of live scleractinian coral cover.

Reef Condition Index

RAP surveys provide an excellent vehicle for rapid documentation of biodiversity of previously unstudied sites. They also afford an opportunity to issue a "report card" on the status or general condition of each reef site. However, this task is problematic. The main challenge is to devise a rating system that is not overly complex, yet accurately reflects the true situation, thus providing a useful tool for comparing all sites for a particular RAP or for comparing sites in different regions. CI's Reef Condition Index (RCI) has evolved by trial and error, and although not yet perfected, shows promise of meeting these goals. Basically, it consists of three equal components: fish diversity, coral diversity, and condition factors.

Fish diversity component—Total species observed at each site. A hypothetical maximum value of 280 species is used to achieve equal weighting. Therefore, the species total from each site is adjusted for equal weighting by multiplying the number of species by 100 and dividing the result by 280.

Coral diversity component—Total species observed at each site. A hypothetical maximum value of 130 species is used to achieve equal weighting. The species total from each site is adjusted for equal weighting by multiplying the number of species by 100 and dividing the result by 130.

Reef condition component—This is the most complex part of the RCI formula and it is therefore instructive to give an example of the data taken from an actual site (1) (Table 2.1):

Table 2.1. Example of data taken from an actual site.

Parameter	1	2	3	4
1. Explosive/Cyanide damage		•		
2. Net damage				•
3. Anchor damage				•
4. Cyclone damage		•		
5. Pollution/Eutrophication				•
6. Coral bleaching			•	
7. Coral pathogens/predators				•
8. Freshwater runoff				•
9. Siltation				•
10. Fishing pressure			•	
11. Coral Cover			•	
Bonus/Penalty Points	-20	-10	+10	+20
Totals	0	-20	+20	+120

Each of 10 threat parameters and the coral cover category (11) is assigned various bonus or penalty points, using a 4-tier system that reflects relative environmental damage: 1. excessive damage (-20 points), 2. moderate damage (-10 points), 3. light damage (+10 points), 4. no damage (+20 points). The impact of fishing pressure (10) is judged from direct observation of fishing activity and is also inferred by the abundance of key target species such as groupers and snappers. Coral cover (11) is rated according to percentage of live hard coral as determined by 100 m line transects (see below): 1. < 26%, 2. 26–50%, 3. 51–75%, 4. 76–100%. In the example shown here the resultant point total is 120. The maximum possible value of 220 (pristine reef with all parameters rated as category 4) is used to achieve equal weighting. The points total for each site is adjusted for equal weighting by multiplying it by 100 and dividing the result by 220. Therefore, for this example the adjusted figure is 54.54.

Calculation of Reef Condition Index—The sum of the adjusted total for each of the three main components described above. Each component contributes one third of the RCI, with a maximum score of 100 for each. Therefore, the top RCI for a totally pristine reef with maximum fish and coral diversity would be 300. Of course, this situation probably does not exist.

Interpretation of RCI values—The interpretative value of CRI will increase with each passing RAP. Thus far, the complete data set contains 104 sites, 47 from the Togean-Banggai Islands and 57 from Milne Bay Province, Papua New Guinea (Allen and Seeto, in press). Table 2.2 provides a general guide to interpretation, based on the data accumulated thus far.

Table 2.2. Interpretation of RCI values based on 104 sites.

General Reef Condition	RCI Value	Percent Of Sites
Extraordinary	>243	4.81
Excellent	214–242	6.73
Good	198–213	28.84
Moderate	170–197	34.61
Poor	141–169	23.08
Very poor	<140	1.92

Coral cover—Data were collected at each site with the use of scuba diving equipment. The main objective was to record the percentage of live scleractinian coral and other major substrates, including dead coral, rubble, sand, soft corals, sponges, and algae. A 100 m measuring tape was used for substrate assessment in three separate depth zones (2–6 m, 8–14 m, and 18–26 m) that varied slightly depending on local conditions. Substrate type was recorded at 1 m intervals along the tape measure, resulting in direct percentages of the

various bottom types for each zone. For the purpose of calculating RCI, the average percentage of coral cover at each site was used (i.e., average for the three transects).

Individual site descriptions

1. Cape Balikapata, Walea Bahi

Time: 1100 hours, dive duration 70 minutes; depth range 2–28 m; visibility approximately 20–25 m; temperature 30°C; slight current. *Site description:* fringing reef with gentle slope to deep water with coral growth to depth of about 30 m then mainly sand bottom; hard coral and rubble dominant substrata; hard coral cover = 62% in 4–6 m, 31% in 12–16 m, 19% in 21–23 m; average hard coral cover 37.3%; relatively high percentage of dead corals and rubble possibly indicative of blast fishing and use of potassium cyanide. RCI = 162.80 (poor).

2. Cape Kedodo, Walea Bahi

Time: 1400 hours, dive duration 75 minutes; depth range 1–30 m; visibility approximately 5 m; temperature 31–32(C; slight current. *Site description:* fringing reef with gentle slope to deep water with coral growth to depth of about 30 m then mainly sand bottom; hard coral dominant substratum; hard coral cover = 61% in 4–6 m, 54% in 12–14 m, 29% in 23–26 m; average hard coral cover 48.0%. RCI = 137.86 (very poor).

3. Western Walea Bahi

Time: 0700 hours, dive duration 80 minutes; depth range 0–25 m; visibility approximately 10–15 m; temperature 30–31°C; slight current. *Site description:* fringing reef with gentle slope to deep water with coral growth to depth of about 30 m then mainly sand bottom; hard coral dominant substratum; hard coral cover = 54% in 4–6 m, 48% in 12–14 m, 29% in 21–22 m; average hard coral cover 43.7%. RCI = 176.26 (moderate).

4. South Walea Kodi

Time: 1015 hours, dive duration 75 minutes; depth range 2–35 m; visibility approximately 15–20 m; temperature 30–31°C; slight current. *Site description:* fringing reef with gentle slope to deep water with coral growth to depth of about 30 m then mainly sand bottom; hard coral dominant substratum; hard coral cover = 31% in 4–6 m, 34% in 12–15 m, 44% in 23–25 m; average hard coral cover 36.3%. RCI = 142.05 (poor).

5. Southwest Walea Kodi

Time: 1300 hours, dive duration 75 minutes; depth range 3–35 m; visibility approximately 15–20 m; temperature 31–32°C; slight current. *Site description:* fringing reef with gentle slope to deep water with coral growth to depth of about 30 m then mainly sand bottom; hard (live) and dead coral dominant substrata; hard coral cover = 16% in 4–5 m, 54% in 12–15 m, 46% in 23–25 m; average hard coral cover 38.7%. RCI = 150.04 (poor).

6.   Reef in front of Vemata Conservation Camp, Malenge Island

Time: 0900 hours, dive duration 80 minutes; depth range 1–28 m; visibility approximately 25 m; temperature 30–31°C; slight current. *Site description:* fringing reef consisting of gentle slope with massive and branching corals on upper part, grading to mainly rubble; dead coral dominant substratum; hard coral cover = 14% in 4 m, 9% in 9 m; average hard coral cover 11.5%; dead coral no doubt partly due to destruction by *Acanthaster planci*—only five individuals observed during transect, but several thousand removed from the area in the months immediately following the RAP survey. RCI = 148.68 (poor).

7.   Kadodo Reef, North Malenge Island

Time: 1200 hours, dive duration 80 minutes; depth range 5–35 m; visibility approximately 20–25 m; temperature 31°C; slight current. *Site description:* north-west extending barrier reef consisting of a gently sloping reef flat and abrupt, nearly vertical wall; hard coral dominant substratum; hard coral cover = 57% in 4–8 m, 60% in 12–14 m, 48% in 23–26 m; average hard coral cover 55.0%; dead corals common at all depths, probably killed by combination of factors including bleaching, cyanide fishing, and *Acanthaster;* abundant sponge growth noted on steep drop-off, particularly in deeper water. RCI = 189.68 (moderate).

8.   Batu Mandi Island, near Kilat Bay

Time: 1600 hours, dive duration 60 minutes; depth range 3–20 m; visibility approximately 10 m; temperature 31–32°C; slight current. *Site description:* fringing reef with relatively poor reef development with considerable dead coral and rubble, and live corals mainly in isolated patches; dead coral and rubble dominant substrata; hard coral cover = 37% in 4 m, 21% in 8 m; average hard coral cover 29.0%. RCI = 143.32 (poor).

9.   North Kadidiri Reef

Time: 0815 hours, dive duration 75 minutes; depth range 3–40 m; visibility approximately 18–20 m; temperature 30–31°C; slight current. *Site description:* barrier reef consisting of a gently sloping reef front and abrupt, nearly vertical wall; hard coral dominant substratum; hard coral cover = 42% in 4–5 m, 56% in 23–25 m; average hard coral cover 49.0%; sponges and encrusting hard corals common on drop-off. RCI = 209.10 (good).

10.   Northwest Kadidiri Reef

Time: 1030 hours, dive duration 75 minutes; depth range 3–25 m; visibility approximately 15 m; temperature 30°C; slight current. *Site description:* barrier reef consisting of a relatively gentle slope; hard coral (*Porites*) dominant substratum; hard coral cover = 52% in 5–7 m, 42% in 10–12 m, 32% in 19–20 m; average hard coral cover 42.0%. RCI = 185.47 (moderate).

11.   Off North Wakai Island

Time: 1400 hours, dive duration 75 minutes; depth range 2–12 m; visibility approximately 4 m; temperature 30–31°C; slight current. *Site description:* gently sloping reef close to mangrove shore, near mouth of Wakai Channel; coral development and diversity generally low due to heavy siltation; dead and live hard coral dominant substrata; hard coral cover = 30 % in 3 m, 35% in 7 m; average hard coral cover 32.5%. RCI = 108.85 (very poor).

12.   Una-una Island (northeast side)

Time: 0815 hours, dive duration 80 minutes; depth range 4–40 m; visibility approximately 15–20 m; temperature 31°C; slight current. *Site description:* fringing reef with steep slope to at least 50 m depth; hard coral dominant substratum; hard coral cover = 53% in 4–6 m, 49% in 11–12 m, 73% in 18–21 m; average hard coral cover 58.3% (highest for Togean Islands); large sponge colonies common on steep slope; very scenic site with good potential as a tourist dive spot. RCI = 196.94 (moderate).

13.   Una-una Island (northeast side)

Time: 1100 hours, dive duration 80 minutes; depth range 2–33 m; visibility approximately 10–15 m; temperature 31°C; slight current. *Site description:* fringing reef with steep slope to at least 50 m depth; hard coral dominant substratum, but percentage cover relatively low due to the steep slope; hard coral cover = 27% in 4–5 m, 37% in 9–13 m, 32% in 19–20 m; average hard coral cover 32.0%; large sponge colonies common on steep slope; very scenic site with potential for recreational dive excursions; extensive rubble at base of slope may suggest past explosive or storm damage; three hanging *bubu* traps noted on transect. RCI = 210.95 (good).

14.   Una Una Island (southwest side)

Time: 1445 hours, dive duration 70 minutes; depth range 2–25 m; visibility approximately 10–15 m; temperature 31°C; slight current. *Site description:* steep-sided, flat-topped (about 300 square m) pinnacle reef about 300 m from shore and coming to within 8–9 m of surface; hard coral dominant substratum; hard coral cover = 58% in 9–12 m, 53% in 21–24 m; average hard coral cover 55.5%; large sponge colonies common on steep slope; this site adjacent to tongue of solidified lava-flow; profuse growth of hard corals on top of pinnacle reef, but reef flat adjacent to coast of island with poor growth, probably hindered by sediment from lava-flow channel (now an intermittent stream bed). RCI = 179.98 (moderate).

15. Pasir-Tengah Atoll

Time: 0730 hours, dive duration 70 minutes; depth range 3–36 m; visibility approximately 10–15 m; temperature 31°C; slight current. *Site description:* atoll environment, with site including steep outer slope, gently sloping foreslope, reef top, and lagoon edge; hard coral dominant substratum; hard coral cover = 58% in 4–5 m, 57% in 9–12 m, 56% in 18–21 m; average hard coral cover 57.0%. RCI = 186.76 (moderate).

16. Western Batudaka Island

Time: 1100 hours, dive duration 80 minutes; depth range 3–50 m; visibility approximately 20 m; temperature 30–31°C; slight current. *Site description:* offshore barrier reef; hard (live) and dead coral dominant substrata; hard coral cover = 62% in 4–5 m, 33% in 11–12 m, 17% in 18–20 m; average hard coral cover 37.3.%; noticeable reduction of live coral and increase in abundance of coralline algae and sponges below 10 m depth. RCI = 199.25 (good).

17. Southwest Batudaka Island

Time: 1545 hours, dive duration 70 minutes; depth range 1–35 m; visibility approximately 10 m; temperature 30–31°C; slight current. *Site description:* barrier/patch reef consisting of steep outer slope on seaward side and gentle sloping lagoon with patch reefs on landward side; hard coral dominant substratum; hard coral cover = 65% in 5–6 m, 57% in 9–10 m, 65% in 18–21 m; average hard coral cover 62.3%. RCI = 168 (moderate).

18. Southwest Batudaka Island

Time: 0745 hours, dive duration 80 minutes; depth range 1–23 m; visibility approximately 5–7 m; temperature 31°C; slight current. *Site description:* coastal fringing reef in silty, sheltered bay; sand dominant substratum; hard coral cover = 38% in 4–5 m, 26% in 13–14 m, 0% below 20 m; average hard coral cover 21.3%. RCI = 174.36 (moderate).

19. Southern Batudaka Island

Time: 1030 hours, dive duration 70 minutes; depth range 3–40 m; visibility approximately 10–15 m; temperature 30–32°C; slight current. *Site description:* barrier/patch reef consisting of steep outer slope on seaward side and gentle sloping lagoon with patch reefs on landward side; hard coral dominant substratum; hard coral cover = 55% in 5–7 m, 47% in 13–15 m, 50% in 20 m; average hard coral cover 50.7%. RCI = 118 (moderate).

20. Southern Batudaka Island

Time: 1345 hours, dive duration 70 minutes; depth range 3–26 m; visibility approximately 7–10 m; temperature 31–33°C; slight current. *Site description:* coastal fringing reef in silty, sheltered bay; hard (live) and dead coral dominant substrata; hard coral cover = 35% in 5–6 m, 37% in 10–12 m, 54% in 20–23 m; average hard coral cover 42.0%. RCI = 173.46 (moderate).

21. Pasir Batang Reef, off Kabalutan Village, Talatakoh Island

Time: 0815 hours, dive duration 80 minutes; depth range 3–37 m; visibility approximately 20 m; temperature 31°C; slight current. *Site description:* coastal fringing reef with shallow reef flat and steep-sided patch reefs in deeper water; hard (live) and dead coral dominant substrata; hard coral cover = 48% in 2–4 m, 47% in 8–10 m, 39% in 18–20 m; average hard coral cover 44.7%. RCI = 181.30 (moderate).

22. Anau Island, off southern Togean Island

Time: 1130 hours, dive duration 80 minutes; depth range 3–25 m; visibility approximately 7–8 m; temperature 30°C; slight current. *Site description:* fringing reef; hard (live) and dead coral dominant substrata; hard coral cover = 53% in 5–6 m, 39% in 8–10 m, 21% in 18–20 m; average hard coral cover 37.7%; relatively high level of siltation, evidently eroded from cultivated areas. RCI = 149.40 (poor).

23. Near southern entrance to Passage between Batudaka and Togean Islands

Time: 1430 hours, dive duration 80 minutes; depth range 3–23 m; visibility approximately 7–8 m; temperature 31–32°C; slight current. *Site description:* fringing reef; dead coral dominant substratum; hard coral cover = 33% in 5–6 m, 29% in 10–11 m, 25% in 20–21 m; average hard coral cover 29.0%; relatively high level of siltation, evidently eroded from cultivated areas—no doubt even greater during rainy season. RCI = 142.42 (poor).

24. Reef between Waleabahi and Talatakoh Islands

Time: 0830 hours, dive duration 80 minutes; depth range 2–34 m; visibility approximately 25–30 m; temperature 30–31°C; slight current. *Site description:* barrier reef several kilometers offshore; hard coral dominant substratum; hard coral cover = 53% in 5–6 m, 59% in 10–11 m, 57% in 19–20 m; average hard coral cover 56.3%. RCI = 209.98 (good).

25. Dondola Island

Time: 1300 hours, dive duration 90 minutes; depth range 3–23 m; visibility approximately 20–30 m; temperature 31°C; slight to moderate currents. *Site description:* fringing reef around isolated small islet with steep outer slope on seaward side and shallow lagoon environment on leeward side, gradually sloping to deep water; hard coral dominant substratum; hard coral cover = 68% in 5–7 m, 59% in 15–17 m, 62% in 20–21 m; average hard coral cover 63.0%; excellent abundance of coral and fishes with clear water and good scenery, an area with excellent potential for a conservation site. RCI = 230.73 (excellent).

26. Cape Pasir Panjang

Time: 0730 hours, dive duration 60 minutes; depth range 3–25 m; visibility approximately 7–8 m; temperature 30–32°C; slight current. *Site description:* large patch reef in coastal bay, mainly on gentle slope from about 8 m to 30 m depth, ending on flat sand bottom; soft corals dominant substratum; hard coral cover = 27% in 4–5 m, 12% in 10–11 m, 23% in 18–20 m; average hard coral cover 20.7%. RCI = 200.22 (moderate).

27. Puludua Island

Time: 1130 hours, dive duration 65 minutes; depth range 1–25 m; visibility approximately 10–15 m; temperature 30–31°C; slight current. *Site description:* coastal fringing reef with shallow reef flat next to shore with narrow spur and groove zone about 50–100 m out, then gradually sloping to depth of about 20 m, sand bottom in 10–20 m with numerous, isolated large coral bommies; soft corals dominant substratum, but mainly confined to shallows (less than 10 m); hard coral cover = 25% in 4–6 m, 36% in 10–11 m, 13% in 20–21 m; average hard coral cover 24.7%. RCI = 190.26 (moderate).

28. Cape Dongolalo

Time: 1340 hours, dive duration 70 minutes; depth range 5–30 m; visibility approximately 10–15 m; temperature 30–31°C; no current at start of dive, but severe, dangerous currents of at least 4–5 knots at end of dive. *Site description:* coastal fringing reef gradually sloping from shore to depth of about 30–35 m, then sand bottom, several rocky islets rising from about 10–15 m depth; soft corals and sand dominant substrata; hard coral cover = 16% in 10–11 m; incredible fish numbers, especially *Odonus niger, Acanthurus mata,* and *Heniochus diphreutes.* RCI = 202.37 (good).

29. Near Lalong Village, North East Peleng Island

Time: 0730 hours, dive duration 65 minutes; depth range 2–25 m; visibility approximately 10–15 m; temperature 30–31°C; no current. *Site description:* coastal fringing reef in sheltered bay, gradually sloping from shore to depth of about 35 m, then sand-rubble bottom; hard corals dominant substratum, very well developed on slope with nearly 100% cover; hard coral cover = 58% in 4–5 m, 91% in 10 m, 96% in 20 m; average hard coral cover 81.7%. RCI = 209.52 (good).

30. Pontil Kecil Island

Time: 1245 hours, dive duration 80 minutes; depth range 2–20 m; visibility approximately 15 m; temperature 30°C; strong current at start, none towards end of dive; *Site description:* fringing reef around small island, extensive (about 50–75 m wide) shallow reef flat with lagoon-like environment, then sloping relatively steeply to about 30–35 m depth; hard corals dominant substratum, very well developed on upper portion of slope; hard coral cover = 53% in 3–4 m, 100% in 10–11 m, 85% in 16–17 m; average hard coral cover 79.3%. RCI = 204.08 (good).

31. Banggai Bay

Time: 1630 hours, dive duration 70 minutes; depth range 2–35 m; visibility approximately 5–8 m; temperature 30°C; no current. *Site description:* extremely sheltered bay with narrow mangrove fringe and steep sandy slope to 30 m depth, then more or less flat; sand dominant substratum; hard coral cover = 0% at all depths; transect on pure sand slope, but later found nearby coral reef along shore in 2–16 m depth with approximately 50% live coral; bottom in 30 m flat and sandy with dense growth of low (about 30 cm high) branching antipatharians. RCI = 167.13 (poor).

32. Bandang Island

Time: 1045 hours, dive duration 70 minutes; depth range 1–40 m; visibility approximately 15 m; temperature 29–30°C; no current in shallows, but moderate in deeper water. *Site description:* fringing reef with broad (about 50–75 m wide) shallow reef flat (dead coral covered with *Padina* alga in 1–2 m), then gradual slope to at least 40 m; good coral growth in shallows to depth of about 10–15 m, then mainly sand-rubble bottom; hard corals dominant substratum; hard coral cover = 87% in 2–4 m, 66% in 10–11 m, 10% in 20–21 m; average hard coral cover 54.3%. RCI = 208.74 (good).

33. East Kenou Island

Time: 1500 hours, dive duration 70 minutes; depth range 2–25 m; visibility approximately 10–15 m; temperature 30°C; no current. *Site description:* fringing reef with broad (about 200 m wide) gradually sloping shallow reef, then relatively steep slope to deep water; hard corals dominant substratum; hard coral cover = 56% in 4–5 m, 34% in 10–11 m, 45% in 18–19 m; average hard coral cover 45.0%; rubble dominant (43%) in transitional zone (10–11 m depth) where slope steepens. RCI = 193.50 (moderate).

34. North Kembonga Island

Time: 0730 hours, dive duration 70 minutes; depth range 1–24 m; visibility approximately 10–12 m; temperature 30°C; no current. *Site description:* fringing reef with broad (about 200–300 m wide) gradually sloping shallow reef, then relatively steep slope to deep water; hard corals, rubble, and soft corals dominant substrata; hard coral cover = 64% in 5–6 m, 25% in 10–11 m, 24% in 19–20 m; average hard coral cover 37.7%; rubble dominant (61%) in transitional zone (10–11 m depth) where slope steepens. RCI = 160.64 (poor).

35. Lagoon between Kembongan and Kokudan islands
Time: 1015 hours, dive duration 70 minutes; depth range 2–25 m; visibility approximately 10–12 m; temperature 30°C; no current. *Site description:* protected lagoon environment between two closely situated (about 1 km apart) high islands, consisting of very broad shallow reef and steep slope profusely covered with live coral to depth of about 25 m, then more or less flat silty sand bottom; hard corals dominant substratum; hard coral cover = 77% in 2–4 m, 87% in 10–11 m, 89% in 20–21 m; average hard coral cover 84.3%. RCI = 142.44 (poor).

36. East Tumbak Island
Time: 1400 hours, dive duration 70 minutes; depth range 1–24 m; visibility approximately 5–15 m; temperature 29–31°C; no current. *Site description:* fringing reef with broad (about 200 m wide) gradually sloping shallow reef, then relatively steep slope to deep water; rubble dominant substratum, especially below 10 m; hard coral cover = 58% in 2–4 m, 19% in 10–11 m, 0% in 12–20 m; average hard coral cover 25.7%. RCI = 163.68 (poor).

37. Atoll between south of Tolopopan Island and east of Silumba Island
Time: 0800 hours, dive duration 65 minutes; depth range 2–24 m; visibility approximately 10–15 m; temperature 31°C; no current. *Site description:* atoll reef with broad (about 200 m wide) gradually sloping shallow reef, then abrupt drop to deep water; hard corals dominant substratum; hard coral cover = 29% in 4–6 m, 41% in 10–11 m, 33% in 18–20 m; average hard coral cover 34.3%. RCI = 193.48 (moderate).

38. Pasibata Reef
Time: 1100 hours, dive duration 70 minutes; depth range 2–24 m; visibility approximately 10–15 m; temperature 30°C; no current. *Site description:* fringing atoll reef; soft corals, hard corals, and sand dominant substrata; hard coral cover = 27% in 2–4 m, 21% in 10–11 m, 19% in 19–20 m; average hard coral cover 22.3%. RCI = 195.70 (moderate).

39. Saloka Island
Time: 1500 hours, dive duration 60 minutes; depth range 4–20 m; visibility approximately 5–10 m; temperature 30–32°C; no current. *Site description:* fringing atoll reef; soft corals and rubble dominant substrata, mostly sand-rubble below 10 m depth; hard coral cover = 5% in 4–5 m, 7% in 10–11 m, 9% in 19–20 m; average hard coral cover 7.0%. RCI = 158.65 (poor).

40. Silula Island
Time: 0730 hours, dive duration 75 minutes; depth range 2–21 m; visibility approximately 10–12 m; temperature 29–30°C; no current. *Site description:* fringing reef; soft corals and rubble dominant substrata, mostly sand-rubble below 10 m depth; hard coral cover = 24% in 2–4 m, 22% in 10–11 m, 9% in 19–20 m; average hard coral cover 18.3%. RCI = 190.41 (moderate).

41. Bangkulu Island (northeastern side)
Time: 1030 hours, dive duration 75 minutes; depth range 3–24 m; visibility approximately 7–8 m; temperature 30°C; no current. *Site description:* fringing reef; sand and rubble dominant substrata; hard coral cover = 14% in 3–4 m, 17% in 10–11 m; average hard coral cover 15.5%. RCI = 173.19 (moderate).

42. Atoll south of Peleng Bay
Time: 1400 hours, dive duration 70 minutes; depth range 2–20 m; visibility approximately 7–10 m; temperature 29–31°C; no current. *Site description:* outer edge of atoll reef; rubble dominant substratum; hard coral cover = 36% in 4–6 m, 14% in 10–11 m, 18% in 19–20 m; average hard coral cover 22.7%. RCI = 185.96 (moderate).

43. West Peleng Island, near Bobo Island
Time: 0730 hours, dive duration 70 minutes; depth range 3–20 m; visibility approximately 5–8 m; temperature 29–30°C; no current. *Site description:* coastal fringing reef in sheltered bay including edge of outer slope, reef flat, and adjacent deep (25 m) lagoon; hard coral and rubble dominant substrata; hard coral cover = 65% in 2–4 m, 35% in 8–10 m, 18% in 19–20 m; average hard coral cover 39.3%. RCI = 149.56 (poor).

44. Patipakaman Peninsula
Time: 1000 hours, dive duration 80 minutes; depth range 2–20 m; visibility approximately 5–8 m; temperature 30°C; no current. *Site description:* coastal fringing reef in sheltered bay including edge of outer slope, reef flat, and adjacent deep (25 m) lagoon; hard coral, rubble, and sand dominant substrata; hard coral cover = 30% in 2–4 m, 22% in 10–11 m, 25% in 17–18 m; average hard coral cover 25.7%. RCI = 161.66 (poor).

45. Dolopo Island (west side)
Time: 1500 hours, dive duration 85 minutes; depth range 4–20 m; visibility approximately 10–15 m; temperature

30°C; no current. *Site description:* barrier reef about 1 km offshore from island, mainly undulating bottom or very slight slope in seaward direction—extensive area of rich coral gardens; hard coral (particularly branching *Acropora* spp.) dominant substrate; hard coral cover = 89% in 3–4 m, 83% in 10–11 m, 83% in 16–17 m; average hard coral cover 85.0% (highest recorded for Banggai Islands). RCI = 207.42 (good).

46. Banyak Islands, West Peleng Island

Time: 0715 hours, dive duration 75 minutes; depth range 2–22 m; visibility approximately 15–17 m; temperature 30°C; no current. *Site description:* fringing reef around island; hard coral dominant substrate, but mainly sandy with isolated corals below 15 m; hard coral cover = 82% in 2–4 m, 74% in 10–11 m, 14% in 17–18 m; average hard coral cover 56.7%. RCI = 239.09 (excellent).

47. Makailu Island

Time: 1030 hours, dive duration 85 minutes; depth range 4–40 m; visibility approximately 20–30 m; temperature 30°C; no current. *Site description:* fringing reef around isolated island lying in deep channel between Peleng Island and mainland Sulawesi, consisting of broad shallow reef flat and nearly vertical drop-off to at least 40–50 m; hard coral dominant substrate; hard coral cover = 58% in 2–4 m, 68% in 12–13 m, 62% in 23–26 m; average hard coral cover 62.7%; large sponge colonies and gorgonian sea fans common on drop-off. RCI = 211.48 (good).

Results

Reef condition—Data used for determining Reef Condition Index is presented in Appendix 2. The hypothetical maximum RCI, as explained previously, is 300. During the cur-

rent survey, values ranged between 137.86 and 238.09. The top 10 sites for reef condition are presented in Table 2.3. Only two sites (25 and 46) were rated as excellent. Most sites were in the poor or moderate categories (Table 2.4).

Coral Cover—It is also useful to consider coral cover independently, as this parameter is often used as an indication of general reef condition, and is widely presented in the literature. A summary of the average hard coral cover at each site is presented in Table 2.5. Detailed data for each of the three depth zones in presented in Appendix 3. The appendix table also includes information on other substrate types (soft corals, sand, rubble, etc.).

A total of 13 sites (27.7%) had an average cover of live scleractinian corals exceeding 50%. Well over half of the sites (27) had an excess of 50% live scleractinian cover in at least one of the three surveyed depth zones. These results compare very favourably with past surveys in the area as well as with other parts of Indonesia. An earlier survey conducted by LIPI (no date given for actual survey, but results published in 1998) at five locations, including the Togean Islands, reported that 29.1% of five reef areas sampled had live hard coral cover in excess of 50%.

Table 2.4. Distribution of relative condition categories based on RCI values.

Relative Condition	No. Sites	% Of Sites
Extraordinary	0	0.00
Excellent	2	4.26
Good	10	21.28
Moderate	19	40.43
Poor	14	29.79
Very Poor	2	4.26

Table 2.3. Top 10 sites for general reef condition.

Site Number	Location	Fish Species	Coral Species	Condition Points	RCI
46	Banyak Islands, Banggai Islands	100	184	210	238.09
25	Dondola Island	76	266	170	230.73
47	Makailu Islands, Banggai Islands	83	197	170	211.48
13	Una-una Islands, Togean Islands	67	230	170	210.95
24	Waleabahi/Talatakoh Islands, Togean Islands	93	184	160	209.98
29	Peleng Is., Banggai Islands	61	188	210	209.52
9	Kadidi Reef, Togean Islands	87	169	180	209.10
32	Bandang Is., Banggai Islands	71	177	200	208.74
45	Dalopo Is., Banggai Islands	89	160	180	207.42
20	Batudaka Is., Togean Islands	83	216	140	204.63

Table 2.5. Average percentage of live coral (for 3 depth zones, except where noted otherwise).

Site	% Cover	Site	% Cover	Site	% Cover
1	37.3	17	62.3	33	45.0
2	48.0	18	32.0*	34	37.6
3	43.6	19	50.6	35	84.3
4	32.7	20	42.0	36	38.5
5	27.6	21	44.6	37	34.3
6	11.5*	22	37.6	38	22.3
7	55.0	23	29.0	39	7.0
8	29.0*	24	56.3	40	18.3
9	52.0	25	63.0	41	15.5*
10	42.0	26	20.6	42	22.6
11	32.5*	27	24.6	43	39.0
12	58.3	28	16.0**	44	25.3
13	32.0	29	81.6	45	85.0
14	37.0	30	79.0	46	56.6
15	57.6	31	0.0	47	62.3
16	36.6	32	54.3		

* two depth zones surveyed; ** one depth zone surveyed.

Comparing the data for the Togean and Banggai Islands reveals that both areas had similar levels of live scleractinian cover, averaging 41.1 and 42.5% respectively. Similarly, 27.7% of Togean reefs had 50% live cover in at least one of the three depth zones surveyed compared to 29.8% of reefs in the Banggai Islands. The Banggai Group boasts the highest average percentage of live scleractinians for a single site (85.0% at site 45), the highest percentage for a single depth zone (100% in 5–19 m at site 30), and the lowest average percentage for a single site (0.0% at Site 30).

Other recent published results for Indonesia include Spermonde Island with 22.6% of sites with more than 50% live hard coral cover (Rachman and Salam, 1993) and Takabonerate with 20.7% of sites exceeding 50% cover (Moka, 1995).

Reef Degradation—Although the live scleractinian coral cover appears to be relatively high throughout the Togean and Banggai groups, there was ample evidence of reef degradation at many survey sites. For example the reef off the southern tip of Walebahi Island (site 1) has clearly been damaged by human activities, particularly the use of explosives for fishing. This was evident from the high percentage of rubble and recently damaged corals present along the survey transects. The death of coral colonies on reefs off southern Walekodi (sites 4–5) is probably due to a high concentration of suspended sediments and the resulting low light penetration. Agricultural development in local rainforest areas has no doubt contributed to soil erosion and consequent sedimentation of local reefs.

We heard dynamite blasts while diving at three sites (1, 33, and 37) and observed damage attributed to dynamite at the majority (86%) of sites in both the Togean and Banggai Islands. In most cases the damage was judged to be moderate to light, but at two sites (41–42) in the Banggai Islands the destruction was severe. We also observed (near site 30) at least one small boat with hookah equipment, which is generally associated with cyanide fishing. Discussions with local villagers in both the Togean and Banggai groups indicate that illegal fishing practices, including the use of dynamite, cyanide and various types of netting, are common.

Coral reefs near Malenge Island (site 6) have been effectively destroyed due to a long history of intensive fishing, including the use of destructive methods such as blast fishing and netting. Fishers are now forced to travel long distances because the local reefs are virtually non-productive.

Bleaching—Coral bleaching is a term used for the sudden loss of symbiotic zooxanthellae harbored by live soft and hard corals. If the condition persists for more than a few days it generally results in the death of the host polyps. The exact causes of bleaching have been the subject of considerable controversy, but there is now overwhelming proof that the condition is accelerated by a prolonged regime of higher than normal sea temperatures, which are commonly associated with El Niño events. Some areas, such as the Maldive Islands, which have lost an estimated 90% of live corals, have been severely affected by the widely publicized 1998 El Niño.

Indonesian reefs appear to have been variably effected by the most recent El Niño. Some areas such as Bunaken Island off Manado, Sulawesi, were severely affected, whereas others, such as the Raja Ampat Islands off Papua Province, have apparently escaped with little or no bleaching reported. During the current survey, bleaching was common at nearly every site in the Togean Islands, with the exception of sites 9, 13, and 18. However, the damage was usually only light, and generally confined to corals in shallow (less than about 10 m) water. By contrast, only a single site (45) in the Banggai Islands was affected by bleaching. The widespread bleaching in the Togean Islands is no doubt correlated with relatively high sea temperatures, which generally ranged from 30–33°C throughout the survey. Although comparative data are lacking for the Togean and Banggai groups, sea temperatures for other areas in Indonesia, Philippines, and Papua New Guinea in which we have worked generally range between 27–29°C.

Coral destruction seen at the Kadoda and Kadidiri reefs (sites 7, 9–10) is of some concern as these are popular dive sites noted for their spectacular underwater scenery, which includes excellent near-vertical walls. We found many dead corals on the reef-flat, as well as in deeper areas, no doubt the result of warm-water induced bleaching.

Coral Predation—Crown-of-thorns starfish (*Acanthaster planci*), a well-documented predator of scleractinian corals, were observed at nine sites in the Togean Islands (2, 4, 5, 6, 8, 9, 10, 12, 23) and four sites in the Banggai Islands (32, 42, 43, and 44). They were usually observed in very low numbers (1–2 per dive), except at Sites 6 and 43, where up to 30 individuals were seen during the course of a single dive.

Conclusions and Recommendations

Compared to other Indonesan areas we have personally observed or that are reported in the literature (Ripang, 1996; Suharsono, 1997; Suharsono et al., 1994; Suripto, 1997) the coral reefs of the Togean and Banggai Islands are in relatively good condition. At most sites we encountered only light to moderate reef damage, primarily due to coral bleaching or illegal fishing methods. However, we did observe some disturbing trends, mainly in the form of widespread illegal fishing methods, that indicate a need for conservation measures in order to arrest what we believe to be a gradual decline in general conditions.

The obvious recommendations involve the establishment of a series of marine reserves throughout the area and effective management of these areas, including proper law enforcement, particularly concerning the apprehension and conviction of anyone engaging in illegal fishing practices. However, carrying out these measures is easier said then done. Judging from unsuccessful efforts by the government in other parts of Indonesia, the chances of implementation are slim at best.

The few conservation success stories in Indonesia emphasize the importance of involving the local community in any initiatives that are undertaken. The community must take pride in its reef resources and learn to effectively manage them. They also need outside support in the form of government-backed effective law enforcement to prevent outsiders from illegally harvesting resources. The community should also be encouraged to become involved in eco-tourism that uses its coral reef resources to attract visitors from other parts of Indonesia and beyond. But any eco-tourism ventures should be community driven with all benefits retained locally by the greater population, rather than lining the pockets of a few Jakarta or Manado-based entrepreneurs. The development of other forms of marine-based income should also be encouraged within local communities in order to relieve general fishing pressure on reef populations. These include "reef-friendly" activities such as seaweed culture and pearl farming. Other recommendations include a formal system of monitoring the general fish catch to facilitate effective management of the reef fishery, and coral rehabilitation projects in areas that have been severely damaged such as in front of the Malenge Island Research Station.

Prospective sites for dive-related eco-tourism include the ten listed in Table 2.3. The area in the vicinity of Dondola Island (site 25) is especially promising as it contains a number of relatively isolated reefs and shoals. If the fauna and reef topography are comparable to that found at Dondola, it would provide a prime location for recreational diving, as well as the establishment of a marine reserve.

Several excellent drop-offs (submarine cliffs) were observed, mainly in the northern portion of the Togean Islands (sites 7, 9, and 16), including Una-una Island (sites 12, 13, and 14), and at Dondola Island (site 25). The only comparable habitat in the Banggais was noted at Makailu Island (site 47). These steep slopes tend to be well populated with an excellent variety of both corals and fishes. Illegal blast fishers generally ignore these areas because of the difficulty in retrieving the catch due to the great depth. Only hook-and-line fishing is effective in such areas and therefore the fishing pressure tends to be relatively low. Consequently, steep drop-offs are prime candidates for conservation sites. Conversely, most dynamite fishing occurs on flat bottoms or gentle slopes in shallow to moderate depths, which facilitates quick and easy harvesting following the blast.

Una-una Island, although considered part of the Togean Group, is well isolated and for conservation purposes probably merits separate consideration. This small volcanic island, measuring about 10 km in diameter, lies across a deep channel, approximately 30 km northwest of Togean Island. The last volcanic eruption occurred in 1986, and although it was consequently evacuated, most of the population has since been resettled. We did not observe any significant reef damage as a result of the eruption, but it is probably restricted to near shore habitats in a few isolated places where former lava flows entered the sea via relatively narrow (up to 100–200 m wide) channels. We observed several of these former lava flows, including one near site 14 on the western side of the island. The northeastern side of the island (sites 12 and 13) harbors some of the most spectacularly scenic coral reefs we encountered during the entire survey.

References

Adhyakso L L. 1997. *Reef Check 197*. Conservation Indonesia. WWF Jakarta.

Allen, G.R., P. Seeto., and T. McGarry Condition of coral reefs in Milne Bay Province, Papua New Guinea. In: G.R. Allen, J. Kinch, and S. A. Mckenna (eds.). *A second Marine Rapid Assessment of Milne Bay Province, Papua New Guinea*. In prep, RAP Bulletin of Biological Assessment, Conservation International, Washington, DC.

COREMAP. 1998. *Coral Reef and Global Change, Adaptation, Acclimation, or Extension, Initial Report of the*

Symposium and Workshop, Boston (2nd), Kalawarta COREMAP, Jakarta.

LIPI. 1998. *Status of Coral Reef in Indonesia, Journalistic Programme Workshop on the Rehabilitation and Management of Coral Reef (COREMAP).* Ilmu Pengetahuan Indonesia, Jakarta.

Moka, W. 1995. *Survey terumbu karang di Taman Nasional Laut Taka Bonerate.* WWF-Universitas Hasanuddin, Ujung Pandang.

Nontji A. 1986. *Indonesian Seas.* Djambatan Press, Jakarta.

Rachman A. and A. Salam. 1993. *Study of the condition of corals to human activities in West Kodya Beach ,Ujung Pandang.* University of Hasanudin, Ujung Pandang.

Ripang, S. 1996. *The Percentage of Coverage and Species Composition of Coral Reef in Barang Lompo Island.* Undergraduate Project. Dept. of Marine Science and Technology, University of Hasanudin, Ujung Pandang.

Soekarno, R. 1989. Comparative studies on the status of Indonesian coral reefs. *Netherlands Journal of Sea Research* 23 (2). Netherlands.

Soekarno, R., M. Hutomo, M.K. Moosa., and P. Darsono. 1983. *Coral Reefs in Indonesia (Its Resources, Problems and Management).* Lembaga Oseanologi Nasional, Lembaga Ilmu Pengetahuan Indonesia, Jakarta.

Suharsono 1997. *A Research Survey on the Biodiversity in Kapposang Island, South Sulawesi.* Pusat Penelitian dan Pengembangan Oseanologi, Ilmu Pengetahuan Indonesia, Jakarta.

Suharsono, A. Budiyanto, N. Hadi, and Gyanto, 1994. *Marine Ecotourism of Three Islands in North Sulawesi (Tundonia, Tenga and Paniki).* Pusat Penelitian dan Pengembangan Oseanologi, Ilmu Pengetahuan Indonesia, Jakarta.

Suharsono, Sukarno, and Siswandono. 1985. Sebaran, keanekaragaman dan kekayaan jenis karang batu di Pulau Kotok Kecil, Pulau-Pulau Seribu. Oseanologi di Indonesia 19. LON-LIPI, Jakarta.

Suripto, U. 1997. *The state of coral reefs in Poto Tano Sumbawa island.* Undergraduate Project, The department of marine science and technology. University of Hasanudin, Ujung Pandang.

Chapter 3

Molluscs of the Gulf of Tomini, Sulawesi, Indonesia

Fred E. Wells

Summary

• A total of 555 species of molluscs belonging to 103 families were collected: 336 gastropods, 211 bivalves, four chitons, two cephalopods, and two scaphopods. Three of the four most diverse families were gastropods: Conidae (63 species), Muricidae (56) and Cypraeidae (41). Veneridae (46) was the most diverse bivalve family. All species for which distributions are known are widespread in the Indo-West Pacific region; none are endemic to Indonesia.

• A total of 31 sites were investigated, with a mean of 65.3 ± 3.5 (SE) species of molluscs collected per site. The richest site was Dondola Island (Site 25) with 115 species and the least diverse was near the southwestern tip of Batudaka Island (Site 17) with 32 species.

• The diversity of molluscs collected on the Togean expedition (555 species) was lower than on the previous CI expeditions to the Calamianes Islands, Philippines (651 species) and Milne Bay, Papua New Guinea (638 species). Diversity in the Togeans was higher than all sites sampled by the author in northwestern Australia except the Montebello Islands, Western Australia (631 species).

• The fauna reported here should not be considered to be the total number of species of molluscs living on coral reefs in the area. The majority present on Indo-West Pacific coral reefs are small and/or cryptic, and no short-term expedition will record all species. To do so would require an intensive survey over a number of years.

• A single octopus was found at one site. This scarcity suggests high fishing pressure, although no octopus fishing was observed. Similarly, low numbers of other target species such as spider shells, conchs, trochus, and abalone were also encountered in the Togeans.

• Aside from the small *Tridacna crocea,* and to a lesser extent *T. squamosa,* very few live giant clams were seen during the expedition. Although various species were recorded from a number of sites, the majority were based on dead shells, and populations appear to have been overfished.

• Low quality seashells are collected worldwide and used for production of trinkets and other products. The specimen shell fishery is relatively small, but some individual shells command high prices. Excessive removal of shells for these purposes has caused substantial damage to coral reef ecosystems in other areas of the Indo-Pacific, particularly the Philippines. Very few specimen shells were found during the expedition to the Togeans, but no fishing for shells was observed.

Introduction

This is the third report of a Rapid Assessment Program (RAP) survey of molluscs on coral reefs in the Indo-Pacific region. The first was undertaken in Milne Bay, Papua New Guinea in October 1997 (Wells, 1998), and the second was to the Calamianes Islands in the Philippines in February 1998 (Wells, 2001).

As indicated in the previous marine RAP reports, molluscs, corals, and fishes were selected as good indicators of overall coral reef biodiversity. Specifically, molluscs were chosen because they have by far the largest diversity of any phylum in the marine environment and the group is relatively well-known taxonomically (at least the larger species). In addition they are ecologically and economically important.

Diversity of molluscs is exceedingly high in tropical waters, particularly in coral reef environments. Gosliner et al. (1996) estimated that approximately 60% of all marine invertebrate species in the Indo-West Pacific region are molluscs. Although molluscan diversity is known to be high on coral reefs, no estimates are available of the total number of species in any Indo-West Pacific coral reef system. In fact, there are few estimates available of the number of molluscs living in any particular area.

In recent years the Western Australian Museum has conducted a number of coral reef surveys in Western Australia and adjoining areas such as Christmas Island and the Cocos (Keeling) Islands, both in the Indian Ocean. Published data from this work provides a basis of comparison with the results of marine RAP surveys.

Methods

Molluscs were collected at 31 sites visited during the survey; unfortunately, prior commitments meant that the author left the expedition before it was completed, and the last 16 sites could not be sampled. Scuba dives were made at all sites examined except site 20. Water depth at this site was too shallow to scuba effectively, and the area was surveyed by snorkel. At all sites as many habitats as possible were collected to provide as complete an indication of diversity as could be made in a short investigation. The tide range in the Togeans is slight, so there is little intertidal area available. However, intertidal areas were sampled where they occurred. The absence of suitable strandlines on the beaches prevented beach collecting of dead shells to obtain additional records; the only substantial beach collection of dead shells was made at site 25.

With the exception of site 25, a single collection was made at each site. Collections were made in deeper water first then the shallows were worked. This process provided a consistent level of sampling effort between sites. The mean time spent at each site was 91 ± 2 minutes (SE) (range 60–120 minutes). Between 80 and 110 minutes were spent at 25 of the 31 sites. Fading light or lack of air restricted collecting times at some sites.

Readily recognizable species were recorded on an underwater slate during each dive. Representatives of all other species were returned to the boat where they were identified using standard shell books, particularly Springsteen and Leobrera (1986). Additional sources available during the trip were: Cernohorsky (1972), Wells and Bryce (1989; 1993), and Gosliner et al. (1996). As there is no book on Indo-Pacific bivalves, and few are included in Springsteen and Leobrera (1986), the Australian reference of Lamprell and Whitehead (1992) was used. Publications on particular

groups were consulted in Perth after the expedition was completed. A small amount of material was taken to the Western Australian Museum for comparison with specimens in the WAM collection.

Results and Discussion

Biodiversity of Molluscs

A total of 555 species of molluscs belonging to 103 families were collected during the expedition. There were 336 gastropods, 211 bivalves, four chitons, two cephalopods, and two scaphopods; no aplacophorans were collected. The preponderance of gastropods is in accord with previous studies using the same techniques, but the proportion of bivalves was slightly higher than on the two previous RAP expeditions. Three of the four most diverse families were gastropods: Conidae (63 species), Muricidae (56) and Cypraeidae (41). Veneridae (46 species) was the most diverse bivalve family, and the second most diverse overall. All of the species for which distributions are known are widespread in the Indo-West Pacific; none are endemic to Indonesia.

A total of 31 sites were investigated, with a mean of 65.3 ± 3.5 (SE) species of molluscs collected per site. A mean of 61.5 ± 3.2 species were collected per site in the Togean Islands. Dondola Island (site 25) was the richest site, with 119 species. However, this was also the only location where specimens were collected along the beach. A total of 85 species were collected during the dive. If only these specimens were counted, site 25 would still be the fifth richest site, with 85 species. site 18 (near the southwest tip of Batudaka Island; 98 species), site 26 (Pasirpanjang Point; 93 species) and Site 6 (east end of Malingi Island; 89 species) had the greatest recorded molluscan diversity. The high species diversity at these sites is correlated with their wider diversity of habitats. Site 17 (near the southwest tip of Batudaka Island), with 32 species had the lowest diversity. Between 43 and 49 species were recorded at a total of eight additional sites. These locations had very low habitat diversity, typically a nearly vertical wall at the edge of the reef that flattened abruptly just below the low tide level. The upper surface of the reef was cemented by coralline algae, and there were no rocks, dead coral, sand, or rubble, all of which normally provide abundant molluscs.

The diversity of molluscs collected on the Togean expedition (555 species) was lower than on previous RAP expeditions: Milne Bay, Papua New Guinea (638 species) and the Calamianes Islands, Philippines (651 species). This is partly a result of the shorter trip to the Togeans, where only 11 days were available for collecting compared to 16 in the Calamianes and 19 in Milne Bay. However, by the end of the Togeans survey, very few additional records were obtained (Figure 3.1 upper).

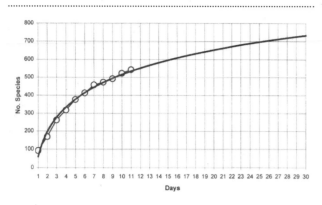

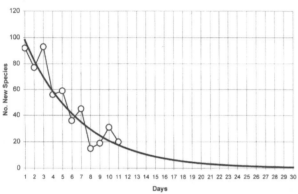

Figure 3.1. Relationships between number of species of molluscs collected and the duration of the expedition. **Upper**—total number of species of molluscs collected during the trip. **Lower**—number of additional species recorded on each day.

After the expedition departed from the Togeans, the number of new records increased as new habitat types were encountered at Dondola Island (Site 25), peninsular mainland sites (26–28), and the Banggai Islands (29–31). It is likely that if molluscs had been collected for the remainder of the expedition the total number of recorded species would have been comparable to Milne Bay and the Calamianes.

During a survey such as this, all species collected on the first day of the expedition are new records. On the second and subsequent days the number of new records declines as progressively more species are recorded for the second or more time. This decrease in new records provides a mechanism for estimating the total number of species of molluscs that would be recorded in the Togeans if additional time were available for the survey (Figure 3.1 lower). This graph suggests that a total of 730 species would be collected in the Togeans using the techniques of this survey if it had continued for a total of 30 days.

The 555 species collected in the Togean-Banggai expedition is greater than the number of species collected on all of the expeditions conducted by the Western Australian Museum in northwestern Australia and adjacent areas except the Montebello Islands, Western Australia, where 631 species were collected (Table 3.1). The Montebellos have a very high degree of habitat complexity that was reflected in all animal groups sampled, not just molluscs. For example, the Montebellos have extensive mangroves and mudflats, neither of which was sampled in the Togeans. Habitat diversity at sites sampled in the Togean expedition was lower, suggesting that there were more species for a given habitat type. In addition, the Montebello expedition had three people actively

Table 3.1. Total mollusc species collected on this trip, as compared with totals for earlier surveys in Papua New Guinea, the Philippines, and Australia.

Location	Collecting Days	Mollusc species	Reference
Togean-Banggai Islands	11	555	This report
Calamianes Group	16	651	Wells, 2000
Milne Bay	19	638	Wells, 1998
Ashmore Reef	12	433	Wells, 1993
Scott/Seringapatam Reef	8	279	Wells & Slack-Smith, 1986
Rowley Shoals	7	260	Wells & Slack-Smith, 1986
Abrolhos Islands	Accumulated data	492	Wells & Bryce, 1997
Montebello Islands	19	631	Wells, Slack-Smith & Bryce, 2001
Cocos (Keeling) Islands	20	380 on survey; total known fauna of 610 species	Wells, 1994
Chagos Islands	Accumulated data	384	Shepherd, 1982
Christmas Is.	12 plus accumulated data	490	Wells & Slack-Smith, 2001
Kimberleys			
1988	19	413	Wells, 1988
1991	19	317	Wells, 1992
1994	13	232	Wells & Bryce, 1995

collecting molluscs during 19 days. These features suggest that total diversity of molluscs in coral reef habitats in the Togeans is probably greater than at the Montebellos.

The data discussed above can be compared to provide information on the *relative* diversity of molluscs in different areas because the same person collected them, with additional help on some expeditions, using the same methodology. However, the 555 species of molluscs recorded from the Togeans should not be considered to be the total number of species to be found on coral reefs in the area.

Diversity of Marine Molluscs in Indonesia

While there is no overall list of the molluscs of Indonesia, two publications indicate the high diversity of the group in the central Indo-West Pacific. The Togean Islands are in northern Indonesia, just south of the Philippine Islands. Just after the turn of the century, Hidalgo (1904–05) recorded 3,121 species of molluscs from the Philippines. This included non-marine molluscs and erroneous records, but indicates the diversity of the fauna at that time. More recently, Springsteen and Leobrera (1986) illustrated nearly 1,700 marine molluscs. Some are deep-water species or are characteristic of non-coral reef shallow habitats such as mangroves. However, Springsteen and Leobrera point out that their book is not intended to be comprehensive and many more species are known to occur in the area. In two popular shell books, Dharma (1988; 1992) illustrated approximately 1,000 of the most common Indonesian shells.

Mollusc Assemblages in the Togean and Banggai Islands

The sites collected during the current expedition can be grouped into five separate geographic areas: the Togean Islands (21 sites); Una-una Island (3 sites); mainland peninsular region (3 sites); the Banggai Islands (3 sites); and Dondola Island (1 site). Figure 2 shows a dendrogram of the relationships between sites using the Bray-Curtis methodology. The dendrogram shows two major clusters in the centre, one on the left with 9 sites and a second cluster on the right with 13 sites. The remaining 9 sites are less closely associated with each other. There is no clear correlation between the four geographic areas sampled and the two clusters. Togean sites occur in both clusters and also among the outlying sites. Sites of the four remaining areas (Una-una Island, peninsular mainland, Banggai Islands, and Dondola Island) are scattered over the dendrogram, though minor clumpings of two sites from the peninsular mainland and the Banggai Islands do occur. The location with the least relationship with the others was site 13 on Una-una Island, where a sandy shoreline with isolated corals was examined for molluscs. The dendrogram suggests that the site clusters reflect habitat types rather than geographical affinities.

The low habitat diversity in the Togean and Banggai Islands was reflected in the relative uniformity of the common species, with a suite of species being found at most sites.

These species are characteristic of shallow water open coral reef systems. The most commonly encountered species was the giant clam *Tridacna squamosa*, which was found at 28 of the 31 sites; however, most of these records were 1–2 live animals or a dead shell. Other species recorded at 20 or more sites were the bivalves *Tridacna crocea, Pedum spondyloidaeum,* and *Barbatia ventricosa* and the gastropods *Rhinoclavis asper, Conus musicus, Coralliophila violacea,* and *Serpulorbis colubrina*. It is interesting that the spider shell, *Lambis lambis,* an edible species which is actively fished in many areas of the Indo-West Pacific, occurred at a large number of sites on both the Calamianes and Milne Bay expeditions; the related species *L. millepedes* was relatively uncommon. This situation was reversed in the Togeans, with *L. millepedes* being recorded at 20 sites and *L. lambis* at only three sites. Another feature of interest was that the China clam, *Hippopus porcellanus,* was only relatively recently (Rosewater, 1982) recognised as a separate species from *H. hippopus*. Both species were found during the Togean expedition, including five sites where the two species were found together.

A number of mollusc species that are common on reef systems in other areas were absent in the Togeans. Although collecting was undertaken on all of the coral reef habitat types available, the suite of species which is characteristic of high energy reef crests was absent. Species such as *Conus ebraeus* and *C. chaldeus* were not recorded. Other species in this habitat, such as *C. imperialis, Cerithium nodulosus* and *Vasum ceramicum* were present as dead shells at 1–2 sites. The lack or scarcity of these species is a real feature rather than a collecting artifact. Weather conditions during the expedition were near perfect—winds were calm and wave action and currents were minimal. Exposed habitats could have been easily collected, but they simply were not present. This suggests that the protected location of the Togean Islands has prevented the development of open, wave swept reef areas.

Drupella cornus and other members of the genus feed actively on corals. Several outbreaks have caused considerable damage on coral reefs, particularly in the Ningaloo Marine Park in Western Australia. Such damage was not observed in the Togeans, with only isolated small patches of corals having been eaten by *Drupella*.

Exploitation of Molluscs in the Togean Islands

A single specimen of octopus was found at one site. Their scarcity is probably a reflection of high fishing pressure, but no octopus fishermen were observed during the survey. This contrasts with the Calamianes expedition, where octopus fishermen were encountered regularly. Similarly, low numbers of other fished species such as spider shells, conchs, and abalone were also encountered on both the Calamianes and Togean expeditions.

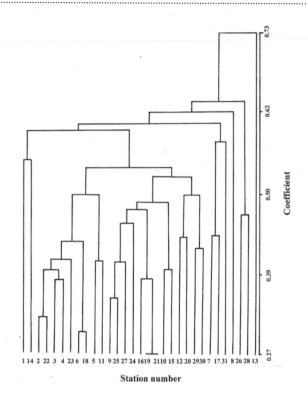

Figure 3.2. Dendrogram showing the relationships between mollusc assemblages at the 31 stations where molluscs were collected.

ing for specimen shells, it is likely that fishermen working on the reefs opportunistically collect them.

Acknowledgments

The diversity of molluscs recorded in the present survey was enhanced by the assistance of the other team members in collecting live and dead shells for me during the survey. I sincerely thank them for this assistance and enjoyed being on the expedition with them. In addition I thank Mr. Andrew Reeves and Dr. Patrick Berry for access to the collections of the Western Australian Museum.

References

Allen, G.R. and R. Steene. 1996. *Indo-Pacific Coral Reef Guide.* Tropical Reef Research, Singapore.

Berry, P .F. (ed.). 1993. *Marine faunal surveys of Ashmore Reef and Cartier Island, North-western Australia.* Records of the Western Australian Museum, Supplement 44.

Cernohorsky, W. O. 1972. *Marine Shells of the Pacific.* Pacific Publications, Sydney, Australia.

Dharma, B. 1988. *Sipit dan kerang Indonesia (Indonesian Shells).* Penerbit: PT Sarana Graha, Jakarta, Indonesia.

Dharma, B. 1992. *Sipit dan kerang Indonesia (Indonesian Shells) II.* Verlag Christa Hemmen, Wiesbaden, Germany.

Gosliner, T. M, D. W. Behrens, and G. C. Williams. 1996. *Coral Reef Animals of the Indo-Pacific.* Sea Challengers, Monterey,California.

Hidalgo, T. 1904–05. *Catalog de los Molluscos Testaceos de las Islas Filipinas, Jolo y Marianas.* Privately published, Madrid.

Lamprell, K. and T. Whitehead. 1992. *Bivalves of Australia. Volume 1.* Crawford House Press, Bathurst, Australia.

Rosewater, J. 1982. A new species of Hippopus (Bivalvia: Tridacnidae). *Nautilus* 96: 3–6.

Sheppard, A. L. S. 1984. The molluscan fauna of Chagos (Indian Ocean) and an analysis of its broad distribution patterns. *Coral Reefs* 3: 43–50.

Springsteen, F.J. and F.M. Leobrera. 1986. *Shells of the Philippines.* Carfel Seashell Museum, Manila, Philippines.

Aside from small *Tridacna crocea,* and to a lesser extent *T. squamosa,* few live giant clams were seen during the expedition. While other *Tridacna* species were recorded from a number of sites, the majority of records were based on dead shells, and populations appear to have been overfished. The margins of many of the Togean reefs are essentially vertical walls that level off to a relatively flat surface one to three meters below the low tide level. The large, colorful giant clams are readily seen in such areas and would be easily harvested.

Low quality seashells are collected worldwide and used for production of trinkets and other products. The fishery for such shells is extremely high in some areas of the Philippines, causing extensive damage to reefs. However, reefs in the Togean and Banggai Islands appear to have suffered little damage from this activity.

The specimen shell fishery is relatively small when compared to the fishery for low quality shells (used mainly for cheap souvenirs and trinkets), but some individual shells can command high prices (in excess of US $1,000) from serious collectors. Very few specimen shells were found during the current survey. While no evidence was found of specific fish-

Wells, F.E. 1988. *Survey of the Invertebrate Fauna of the Kimberley Islands.* Unpublished Report, Western Australian Museum, Perth, pp. 1–51.

Wells, F.E. 1992. Part IV. Molluscs. pp. 30–42. In: G. J. Morgan (ed.). *Survey of the Aquatic Fauna of the Kimberley Islands and Reefs,* Western Australia. Western Australian Museum, Perth.

Wells, F.E. 1993. Molluscs of Ashmore Reef and Cartier Island. pp. 25–45. In: P. F. Berry (ed.) *Marine faunal surveys of Ashmore Reef and Cartier Island, North-western Australia.* Records of the Western Australian Museum, Supplement 24: 25–45.

Wells, F.E. 1994. Marine Molluscs of the Cocos (Keeling) Islands. *Atoll Research Bulletin* 410: 1–22.

Wells, F.E. 1998. Part 3. Molluscs of Milne Bay Province, Papua New Guinea. In: T. B. Werner and G.R. Allen (eds.). *A Rapid Biodiversity Assessment of the Coral Reefs of Milne Bay Province, Papua New Guinea.* RAP Working Papers 11, Conservation International, Washington, DC.

Wells, F.E. 2001. Part 2. Molluscs of the Calamianes Islands, Palawan Province, Philippines. In: T. B. Werner and G.R. Allen (eds.). *A Rapid Marine Biodiversity Assessment of the Calamianes Islands, Palawan Province, Philippines.* Bulletin of the Rapid Assessment Program 17, Conservation International, Washington, DC.

Wells, F.E. and C.W. Bryce. 1989. *Seashells of Western Australia.* Western Australian Museum, Perth.

Wells, F.E. and C.W. Bryce. 1993. *Seaslugs of Western Australia.* Western Australian Museum, Perth.

Wells, F.E. and C.W. Bryce. 1995. Molluscs. In: F.E. Wells, R. Hanley, and D. I. Walker. 1995. *Survey of the Marine Biota of the Southern Kimberley Islands, Western Australia.* MS report to the National Estates Grant Programme. Western Australian Museum, Perth.

Wells, F.E. and C.W. Bryce. 1997. A preliminary checklist of the marine macromolluscs of the Houtman Abrolhos Islands, Western Australia. pp. 362–384. In: F.E. Wells (ed). *Proceedings of the Seventh International Marine Biological Workshop: The Marine Flora and Fauna of the Houtman Abrolhos Islands, Western Australia.* Western Australian Museum, Perth.

Wells, F.E. and S.M. Slack-Smith. 1986. Part IV. Molluscs. In: P. F. Berry (ed.) *Faunal Survey of the Rowley Shoals and Scott Reef, Western Australia.* Records of the Western Australian Museum, Supplement 25: 41–58.

Wells, F.E. and S.M. Slack-Smith. 2001. Molluscs of Christmas Island. In: P. F. Berry and F.E. Wells (eds.). *Survey of the Marine Fauna of the Montebello Islands, Western Australia and Christmas Island, Indian Ocean.* Records of the Western Australian Museum, Supplement 59: 103–115.

Wells, F.E., S.M. Slack-Smith, and C.W. Bryce. 2001. Molluscs of the Montebello Islands. In: P. F. Berry and F.E. Wells (eds.). *Survey of the Marine Fauna of the Montebello Islands, Western Australia and Christmas Island, Indian Ocean.* Records of the Western Australian Museum, Supplement 59: 29–46.

Chapter 4

Reef Fishes of the Togean and Banggai Islands, Sulawesi, Indonesia

Gerald R. Allen

Summary

- A list of fishes was compiled for 47 sites, including 24 in the Togean Islands, 19 in the Banggai Islands and four on the shoals and mainland peninsula that separates the two groups. The survey involved 72 hours of scuba diving by G.R. Allen to a maximum depth of 50 m.

- The Togean and Banggai Islands have a diverse reef fish fauna. A total of 819 species were observed or collected during the present survey. An extrapolation method using six key index families (Chaetodontidae, Pomacanthidae, Pomacentridae, Labridae, Scaridae, and Acanthuridae) indicates a total fauna consisting of at least 1,023 species.

- Species numbers at visually sampled sites ranged from 70 to 266, with an average of 173 per site. Banggai sites (average 176 species) were generally more diverse than Togean sites (average 166 species). The 4 sites at shoals and the mainland peninsula separating the two groups had an average of 201 species.

- Despite a greater sampling effort (24 versus 19 sites) in the Togeans, more species were recorded from the Banggai Islands. The respective estimated total reef faunas for these areas derived from the Coral Fish Diversity Index (CFDI) regression formula are 799 and 908.

- The fish fauna of the Togean and Banggai Islands consists mainly of species associated with coral reefs, which have relatively broad distributions in the Indo-Pacific region. Atolls were the richest of the major habitats with 200 species per site, while sheltered fringing reefs were the poorest with 145.1 species per site.

- Damselfishes (Pomacentridae), wrasses (Labridae), and gobies (Gobiidae) are the dominant families in the Togean and Banggai Islands in terms of number of species (99, 97, and 85 respectively) and number of individuals.

- The overwhelming majority of reef fishes in the Togean and Banggai Islands are either carnivores or planktivores, feeding on a wide variety of invertebrates and fishes. The remaining 26% of the fauna are either herbivorous or omnivorous.

- A total of seven undescribed species were collected during the survey including five wrasses, a damselfish, and a blenny (*Ecsenius;* family Blenniidae).

- Due to its extremely restricted distribution, low reproductive rate, and high level of harvesting for the aquarium trade, the survival of the endemic Banggai Cardinalfish (*Peterapogon kauderni*) is seriously threatened.

Introduction

This section of the report contains comprehensive documentation of the reef and shore fish fauna of the Togean and Banggai Islands based on results of Conservation International's RAP during October–November 1998. The background of this project and description of the 47 survey sites are provided elsewhere in this report.

The principle aim of the fish survey was to provide a comprehensive inventory of the reef-associated species. This segment of the fauna includes fishes living on or near coral reefs down to the limit of safe sport diving or approximately 45 m depth (although observations may extend to 50–60 m depth or more if visibility is good). Survey activities therefore excluded deepwater fishes and offshore pelagic species such as flyingfishes, tunas, and billfishes.

The results of this survey facilitate a comparison of the faunal richness of the Togean and Banggai Islands with other parts of South East Asia and adjoining regions. However, the list of fishes presented is still incomplete due to the time restriction of the survey (16.5 days) and the secretive nature of many small reef species. Nevertheless, a basic knowledge of the cryptic component of the fauna in other areas and an extrapolation method that uses key "index" families can accurately estimate the overall species total.

Methods

The fish portion of this survey involved 72 hours of scuba diving by G.R. Allen to a maximum depth of 50 m. A list of fishes was compiled for 47 sites. The basic method consisted of underwater observations, in most cases during a single, 60–90 minute dive at each site. The name of each species encountered was written on a plastic sheet attached to a clipboard. The technique usually involved a rapid descent to 30–50 m, then a slow, zigzag ascent path back to the shallows. The majority of time was spent in the 2–12 m depth zone, which consistently harbors the largest number of species. The visual transect at each site included a representative sample of all available bottom types and habitat situations, for example adjacent mangroves and seagrass beds, shallow reef flats, steep drop-offs, caves (using a flashlight if necessary), rubble and sand patches, etc. Only the names of fishes for which identification was absolutely certain were recorded. However, there were very few, less than about 2% of those observed, which could not be identified to species level.

The visual survey was supplemented with 10 small collections procured with the use of the ichthyocide rotenone and several specimens that were collected with a rubber-sling propelled multi-prong spear. The purpose of the rotenone collections was to flush out small crevice and sand-dwelling fishes (e.g., eels and tiny gobies) that are never recorded with the visual technique. A total of 52 species were added by using this method.

Results

A total of 819 species belonging to 273 genera and 75 families were recorded during the survey (Appendix 5). Nearly all of the fishes appearing in the list are illustrated in Allen (1991, 1993, and 1997), Myers (1989), Kuiter (1992), or Randall et al. (1990).

General faunal composition

The fish fauna consists mainly of species associated with coral reefs. The most abundant families in terms of number of species are damselfishes (Pomacentridae), wrasses (Labridae), gobies (Gobiidae) cardinalfishes (Apogonidae), groupers (Serranidae), butterflyfishes (Chaetodontidae), blennies (Blenniidae), surgeonfishes (Acanthuridae), parrotfishes (Scaridae), and snappers (Lutjanidae). These 10 families collectively account for about 65.8% of the total observed fauna (Table 4.1).

The relative abundance of fish families in this region is similar to other reef areas in the Indo-Pacific, although the ranking of individual families is variable, as shown in Table 4.2.

Table 4.1. The most abundant families of fish in terms of number of species.

Rank	Family	Species	% of Total Species
1	Pomacentridae	99	12.1
2	Labridae	97	11.8
3	Gobiidae	85	10.4
4	Apogonidae	57	7.0
5	Serranidae	46	5.6
6	Chaetodontidae	38	4.6
7	Blenniidae	35	4.3
8	Acanthuridae	33	4.0
9	Scaridae	26	3.2
10	Lutjanidae	23	2.8

Habitats and fish biodiversity

The species occurring at an individual locality are largely dependent on the availability of shelter and food. Coral and rocky reefs exposed to periodic strong currents are by far the richest habitat in terms of fish biodiversity. These reefs provide an abundance of shelter for fishes of all sizes, and the currents are vital for supporting numerous planktivores, the smaller of which provide food for larger predators. The highest number of fish species was usually recorded at sites with the following features: (1) predominantly coral or rock reef substratum, (2) relatively clear water, (3) periodic strong currents, and (4) presence of additional habitats (sand-rubble, seagrass, mangroves, etc.) in close proximity (i.e., within easy swimming distance of the primary coral reef habitat). The number of species found at each site is shown in Table 4.3. The total number ranged from 70 to 266, with an average of 173 per site. Banggai sites (Sites 29–47: average 176 species) were generally more diverse than Togean sites (Sites 1–24: average 166 species). The four sites (25–28) at shoals and the mainland peninsula separating the two groups had an average of 201 species.

Types of substrata

The best sites for fishes (Table 4.4) were usually locations where live coral was a dominant feature of the seascape, although there was usually a mixture of other bottom types, particularly sand or rubble. Mangroves, seagrass beds, and pure sand-rubble areas were the poorest areas for fish diversity. Silty bays and harbor, although sometimes supporting a variety of hard corals, also had impoverished fish communities. The five sites (2, 4, 11, 35, 44) where less than 140 species were recorded invariably consisted of sheltered locations with reduced visibility and substantial siltation.

Tables 4.5 and 4.6 present a comparison of the fish fauna of the major areas and habitats that were surveyed. The shoals and mainland peninsula that separate the Togean and Banggai Islands exhibited the most species (201.3 per site), whereas the Togean Islands (162.1 per site) had the fewest. Atolls were the richest of the major habitats with 200 species per site, while sheltered fringing reefs were the poorest with 145.1 species per site.

Index of fish diversity (CFDI)

Due to the need for a convenient method of assessing and comparing overall coral reef fish diversity in the Indo-Pacific

Table 4.2. Family rankings in terms of number of species for various localities in the Indo-Pacific region. Data are from unpublished survey results by the author except Milne Bay and the Chagos Islands which are from Allen (1998) and Winterbottom et al. (1989), respectively.

Family	Togean and Banggai Islands	Calamianes Group, Philippines	Milne Bay, PNG	Flores, Indonesia	Komodo, Indonesia	Chagos Islands
Pomacentridae	1st	2nd	1st	3rd	1st	3rd
Labridae	2nd	3rd	3rd	2nd	2nd	2nd
Gobiidae	3rd	1st	2nd	1st	3rd	1st
Apogonidae	4th	4th	4th	4th	4th	6th
Serranidae	5th	7th	4th	5th	5th	4th
Chaetodontidae	6th	5th	7th	7th	6th	11th
Blenniidae	7th	6th	6th	6th	8th	9th
Acanthuridae	8th	9th	8th	8th	7th	8th
Scaridae	9th	8th	10th	10th	10th	12th
Lutjanidae	10th	10th	9th	9th	9th	7th

Table 4.3. Number of species observed at each site during a survey of the Togean and Banggai Islands.

Site	Species	Site	Species	Site	Species
1	178	17	162	33	215
2	134	18	149	34	151
3	154	19	163	35	120
4	133	20	216	36	174
5	155	21	174	37	213
6	179	22	106	38	210
7	181	23	136	39	175
8	145	24	184	40	195
9	169	25	266	41	183
10	208	26	170	42	189
11	70	27	193	43	147
12	161	28	176	44	138
13	230	29	188	45	160
14	175	30	181	46	184
15	202	31	139	47	197
16	208	32	177		

Table 4.4. Ten richest fish sites during a survey of the Togean-Banggai Islands.

Site Number	Location	Total Fish Species
25	Dondola Is.	266
13	N tip of Una-una Is., Togean Islands	230
33	SE end Kenau Is., Banggai Islands	215
37	Atoll S of Treko Islands, Banggai Islands	213
38	E side Pulau Pasibata reef, Banggai Islands	210
10	S Kadidi barrier reef, Togean Islands	208
16	W end Batudaka barrier reef, Togean Islands	208
47	Makailu Is., Banggai Islands	197
40	S side Sidula Is., Banggai, Islands	195
42	Atoll off NW Bangkulu Is., Banggai Islands	189

Table 4.5. Average number of fish species recorded for major areas.

Area	Average Number Species/Site
Shoals and mainland peninsula (Sites 25–28)	201.3
Una-una Is. (Sites 12–14)	188.6
Banggai Islands (Sites 29–47)	175.6
Main Togean Islands (Sites 1–24, except 12–14)	162.1

Table 4.6. Average number of fish species recorded for major habitats.

Habitat	Average Number of Species
Atolls (4 sites)	200.0
Platform reefs, pinnacle reefs, and offshore cays (10 sites)	192.3
Exposed fringing reefs (11 sites)	178.5
Barrier reefs (8 sites)	173.8
Sheltered fringing reefs (14 sites)	145.1

region, I have devised a rating system based on the number of species present belonging to the following six families: Chaetodontidae, Pomacanthidae, Pomacentridae, Labridae, Scaridae, and Acanthuridae. These families are particularly good indicators of fish diversity for the following reasons:

- They are taxonomically well-documented.

- They are conspicuous diurnal fishes that are relatively easy to identify underwater.

- They include the "mainstream" species, which truly characterize the fauna of a particular reef. Collectively, they usually comprise more than 50% of the observable fishes.

- The families, with the exception of Pomacanthidae, are consistently among the 10 most speciose groups of reef fishes inhabiting a particular locality in the tropical Indo-west Pacific region.

- Labridae and Pomacentridae in particular are very speciose and use a wide range of associated habitats in addition to coral-rich areas.

The method of assessment consists simply of counting the total number of species present in each of the six families. It is applicable at several levels:

- single dive sites;

- relatively restricted localities such as the Togean-Banggai Islands;

- countries, major island groups, or large regions.

Coral Fish Diversity Index (CFDI) values can be used to make a reasonably accurate estimate of the total coral reef fish fauna of a particular locality by means of a regression formula. This feature is particularly useful for large regions, such as Indonesia and the Philippines, where reliable totals are lacking. Moreover, the CFDI predictor value can be used to gauge the thoroughness of a particular short-term survey that is either currently in progress or already completed. For example, due to time restraints and heavy reliance on visual observations, 819 species were recorded during the present survey. However, according to the CFDI predictor formula an approximate total of 1,100 species could be expected, revealing that 74% of the fauna was actually surveyed.

The above-mentioned regression formula was obtained from an analysis of 35 Indo-Pacific locations that have been comprehensively studied and for which reliable species lists exist. The data were first divided into two groups: those from relatively restricted localities (surrounding seas encompassing less than 2,000 km^2) and those from much larger areas (surrounding seas encompassing more than 50,000 km^2). Simple regression analysis revealed a highly significant difference ($P = 0.0001$) between these two groups. Therefore, the data were separated and subjected to an additional analysis. The Macintosh program Statview was used to perform simple linear regression analyses on each data set in order to determine a predictor formula, using CFDI as the predictor variable (x) for estimating the independent variable (y) or total coral reef fish fauna. The resultant formulae were obtained: 1. total fauna of areas with surrounding seas encompassing more than 50,000 km^2 = 4.234(CFDI) - 114.446 (d.f. = 15; R2 = 0.96; P = 0.0001); 2. total fauna of areas with surrounding seas encompassing less than 2,000 km^2 = 3.39 (CFDI) - 20.595 (d.f. = 18; R2 = 0.96; P = 0.0001).

Table 4.7. Interpretation of Coral Fish Diversity Index (CFDI) values in terms of relative categories of biodiversity.

Relative Biodiversity	CFDI Values		
	Single Site	Restricted Area	Country-Region
Extraordinary	>150	>330	>400
Excellent	130–149	260–329	330–399
Good	100–129	200–259	220–329
Moderate	70–99	140–199	160–219
Poor	40–69	50–139	80–159
Very Poor	<40	<50	<80

CFDI values obtained for individual sites, relatively restricted areas (i.e. Togean and Banggai Islands), or larger regions or countries can be readily interpreted by referring to Table 4.7, which is based on numerous surveys in the Indo-Pacific by the author and various colleagues.

Summary of Coral Fish Diversity Index (CFDI) Assessment

A selection of CFDI values for individual dive sites in the "Coral Triangle," including several from the current survey, are compared in Table 4.8. Only one site (25) was ranked in the extraordinary biodiversity category and none were rated excellent. Most sites were assessed as being moderate to good compared to other areas in the Indo-Pacific.

An overall CFDI total of 308 was recorded for the Togean and Banggai Islands, which compares favorably with other restricted Indo-Pacific localities, being ranked third overall (Table 4.9).

Using CFDI values to compare more extensive regions, it can be seen from Table 4.10 that Indonesia possesses the world's richest reef fish fauna with an estimated species total of 2,056 species, of which approximately 50% occur within the combined Togean-Banggai Islands region.

Table 4.8. Coral Fish Diversity Index (CFDI) values for selected single dive transects undertaken by the author at various localities. Those resulting from the present survey are in bold.

Transect Site	CFDI	% Total Spp.	Total Spp.
Boirama Is., Milne Bay Prov., PNG	160	59.3	270
Wahoo Reef, Milne Bay Prov., PNG	159	64.9	245
Dondola Is., E. Sulawesi	158	59.4	266
Kri Is., Raja Ampat Is., Papua Province, Indonesia	156	57.1	273
Irai Is., Milne Bay Prov., PNG	156	58.2	268
Seraja Besar, W. Flores, Indonesia	136	64.4	211
Tara Is., Calamianes Is., Philippines	132	63.5	208
SE end Kenau Is., Banggai Is.	132	61.4	215
Pulau Besar, Maumere, Flores	130	54.4	239
Dimipac Is., Calamianes Is., Philippines	129	63.2	204
Tandah Putih, Peleng Is., Banggai Is.	126	67.0	188
SW Tara Is., Calamianes Is., Philippines	126	64.3	196
Una-una Is., Togean Is.	125	54.3	230
N. Komodo Is., Indonesia	122	60.4	202
Kimbe Bay, W. New Britain, PNG	120	64.9	185
Halsey Harbour, Culion Is., Philippines	119	56.4	211
Padoz Reef, Madang, PNG	111	56.3	197
Tripod Reef, Madang, PNG	105	63.6	165
Pig I. Lagoon, Madang, PNG	102	58.0	176
Jais Aben Reef, Madang, PNG	94	57.3	164
E. Rinca Is., Indonesia	61	50.8	120
CRI Reef, Madang, PNG	58	43.9	132
Kimbe Bay, W. New Britain, PNG	57	63.3	90
Pulau Sedona, Bintan Is., Riau Islands	41	51.9	79
Wakai, NE Batudaka Is., Togean Is.	36	51.4	70

Table 4.9. Coral Fish Diversity Index (CFDI) values for restricted localities, number of coral reef fish species as determined by surveys to date, and estimated numbers using the CFDI regression formula (refer to text for details).

Locality Site	CFDI	No. Reef Fishes	Estimated Reef Fishes
Togean and Banggai Is., Indonesia	**308**	**819**	**1023**
Maumere Bay, Flores, Indonesia	333	1111	1107
Milne Bay, Papua New Guinea	318	1084	1057
Komodo Islands, Indonesia	280	722	928
Madang, Papua New Guinea	257	787	850
Kimbe Bay, Papua New Guinea	254	687	840
Manado, Sulawesi, Indonesia	249	624	823
Capricorn Group, Great Barrier Reef	232	803	765
Ashmore/Cartier Reefs, Timor Sea	225	669	742
Kashiwa-Jima Is., Japan	224	768	738
Scott/Seringapatam Reefs, W. Australia	220	593	725
Samoa Islands	211	852	694
Chesterfield Islands, Coral Sea	210	699	691
Sangalakki Is., Kalimantan,	201	461	660
Bodgaya Islands, Sabah, Malaysia	197	516	647
Izu Islands	190	464	623
Christmas Is., Indian Ocean	185	560	606
Sipadan Is., Sabah, Malaysia	184	492	603
Rowley Shoals, W. Australia	176	505	576
Cocos-Keeling Atoll, Indian Ocean	167	528	545
North-West Cape, W. Australia	164	527	535
Tunku Abdul Rahman Is., Sabah	139	357	450
Lord Howe Is., Australia	139	395	450
Monte Bello Islands, W. Australia	119	447	382
Bintan Is., Indonesia	97	304	308
Kimberley Coast, W. Australia	89	367	281
Cassini Is., W. Australia	78	249	243
Johnston Is., Central Pacific	78	227	243
Midway Atoll	77	250	240
Rapa	77	209	240
Norfolk Is.	72	220	223

Table 4.10. Coral Fish Diversity Index (CFDI) values for regions or countries with figures for total reef and shore fish fauna (if known), and estimated fauna based on CFDI regression formula.

Locality	CFDI	No. Reef Fishes	Estimated Reef Fishes
Indonesia	501	2056	2032
Australia (tropical)	401	1714	1584
Philippines	387	?	1525
Papua New Guinea	362	1494	1419
S. Japanese Archipelago	348	1315	1359
Great Barrier Reef, Australia	343	1325	1338
Taiwan	319	1172	1237
Micronesia	315	1170	1220
New Caledonia	300	1097	1156
Sabah, Malaysia	274	840	1046
Northwest Shelf, Western Australia	273	932	1042
Mariana Islands	222	848	826
Marshall Islands	221	795	822
Ogasawara Islands, Japan	212	745	784
French Polynesia	205	730	754
Maldive Islands	219	894	813
Seychelles	188	765	682
Society Islands	160	560	563
Tuamotu Islands	144	389	496
Hawaiian Islands	121	435	398
Marquesas Islands	90	331	267

Table 4.11. Number of damselfish species at selected localities.

Locality	Number of Species
Indonesia	152
Philippines	122
New Guinea	109
Komodo Indonesia	100
Togean-Banggai Islands	**99**
Calamianes Group Philippines	97
Milne Bay PNG	97
N. Australia	95
W. Thailand	60
Fiji Islands	60
Bintan Is., Indonesia	48
Maldives	43
Red Sea	34
Society Islands	30
Hawaii	15

Zoogeographic affinities of the fish fauna

The reef fishes of the Togean and Banggai Islands, and Sulawesi in general, belong to the overall Indo-West Pacific faunal community. They are very similar to those inhabiting other areas within this vast region, stretching eastward from East Africa and the Red Sea to the islands of Micronesia and Polynesia. Although most families, and many genera and species, are consistently present across the region, the species composition varies greatly according to locality.

The Togean and Banggai Islands are an integral part of the Indo-Australian Archipelago, the richest faunal province on the globe in terms of biodiversity. The region forms the center of what is sometimes referred to as the Coral Triangle. Species richness generally declines with increased distance from the Indonesian center of this region, although a secondary region of speciation in the Red Sea-Western Indian Ocean counters this effect. The damselfish family Pomacentridae provides an excellent, very typical, example of this phenomenon (Table 4.11). Approximately 152 species occur in the Indonesian Archipelago, 122 in the Philippines, 109 in New Guinea, and only 15 and 16 respectively at Hawaii and Pitcairn Island, which lie on the outermost fringe of the region (Allen, 1991). The total of 99 species in an area the size of the Togean and Banggai Islands is indicative of an overall rich reef-fish fauna.

Behavioral modes and feeding relationships

The majority of the area's fishes are diurnal coral reef species that either live or forage on or near the bottom (Fig. 4.1). Approximately 10% of the species are nocturnally active. One shortcoming of the visual survey technique is that cryptic species, living either in caves and fissures or beneath the substratum, are not comprehensively sampled. The author's extensive data indicates that the cryptic component accounts for an average of 24% of the total fauna.

The association of consumers and consumed organisms, involving an incredibly diverse array of plants, invertebrates, and vertebrates, results in a complex, highly intertwined food-web. The overwhelming majority of Togean-Banggai fishes are carnivores and/or planktivores, feeding on a wide range of invertebrates and fishes (Fig. 4.2). About 26% of the Calamianes species are either herbivorous or omnivorous. This breakdown is typical for coral reef fish communities in general. Detailed information on the feeding habits of individual species is provided in the annotations in Appendix 1.

Habitats and fish biodiversity

The Togeans-Banggai area has a rich marine icthyofauna in comparison with other coral reef areas in the Indo-Pacific region. It is mainly composed of widely distributed elements that are recruited as postlarvae, after a variable pelagic stage. The total species present at a particular locality is ultimately dependent on the availability of food and shelter and the

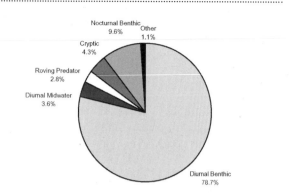

Figure 4.1. Activity modes of Togean and Banggai fishes.

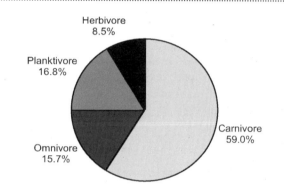

Figure 4.2. Feeding modes of Togean and Banggai fishes.

diversity of substrata. Coral reef habitats with a rich assemblage of hard and soft corals, interspersed with sections of sand, rubble, and seaweed generally harbor the richest fish communities.

Most reef fishes found in this area have relatively widespread distributions within the Indo-Pacific region. Nearly all coral reef fishes have a pelagic larval stage of variable duration, depending on the species. Therefore, the dispersal capabilities and length of larval life of a given species is usually reflected in the geographic distribution. The main zoogeographic categories for Togean-Banggai fishes are presented in Figure 4.3. The largest segment of the fauna consists of species that are broadly distributed in the Indo-West and Central Pacific region from East Africa to the islands of Oceania. The remaining species have more restricted distributions within the Indo-Pacific region or occur circum tropically.

Comparison between fish faunas of the Togean and Banggai Islands
A comparison of the fish faunas of the two main island groups that were surveyed is summarized in Table 4.12. Despite a greater sampling effort (24 versus 19 sites) in the Togeans, more species were recorded from the Banggai Islands. The CFDI totals were 242 for the Togean Islands and 274 for the Banggai Islands. The respective estimated total reef faunas for these areas derived from the CFDI regression formula are 799 and 908.

It is difficult to assess real differences between the two areas on the basis of such a brief survey. Certainly most of the species that were found in one area and not the other, and which were based on just a few sightings, could be expected to occur in both areas given a longer sampling period. Also, there was a disproportionate sampling of various habitats correlated with geography. For example mangroves,

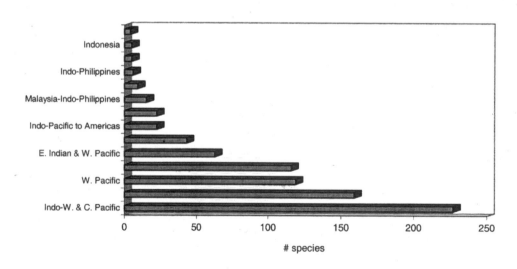

Figure 4.3. Zoogeographic analysis of Togean-Banggai fishes.

which bordered a reef area (sites 3, 4, and 20) were surveyed in the Togeans, but not the Banggais. This accounts for several mangrove-associated fishes being recorded only at the former locality. Likewise, precipitous outer reef drop-offs were well represented in the Togeans (sites 7, 9, 12, 13–16, and 24), but present at only one locality in the Banggais (site 47). In spite of the sampling and habitat bias there appear to be genuine faunal differences between the two island groups. Table 4.13 presents a list of species that appeared to be genuinely restricted to one or the other of these groups. This list includes obvious, easily-recognizable species that were relatively common at one location, but absent at the other.

Species of special interest

A total of seven undescribed species were collected during the survey. These discoveries emphasize the still incomplete knowledge of Indonesia's reef fish fauna. The status and distribution of the new species is discussed below.

Amblypomacentrus clarus (Pomacentridae)—Four specimens, 25–42 mm SL, were collected from a bare sandy slope in Banggai Harbour (site 31) at a depth of 16–18 m. Remarkably, young specimens were seen sheltering in an active jawfish burrow. Comparisons with the only known species in the genus (*A. breviceps*), were readily facilitated by the presence of both adults and juveniles of each species at the same site. This species was described by Allen and Adrim (2000) on the basis of the RAP specimens. It was also recently photographed in Bali by J. Randall of the Bishop Museum, Hawaii.

Choerodon sp. (Labridae)—A single 83 mm SL female specimen was speared in 30 m depth off Bandang Island, near Banggai Island (site 32). The species was common (approximately 50 seen) over a rubble bottom. It will be described by the author and J. Randall. The species is also known from the Coral Sea.

Cirrhilabrus aurantifasciatus (Labridae)—An 84 mm SL male was speared in 12 m depth at Batudaka Island, Togean Islands (site 20). This species, which has a purple body and bright orange back, was common in the Togeans, particularly in rubble areas. It has also been observed on the Sulawesi mainland portion of the Gulf of Tomini, east of Gorontalo, and is known from the Komodo area. It was recently described by Allen and Kuiter (1999).

Cirrhilabrus tonozukai (Labridae)—Three specimens, 56–58 mm SL, were speared in 20–35 m depth at Kenau and Tumbak islands (sites 33 and 36) in the Banggai Group. This spectacular species is related to *C. filamentosus* from Java, Bali, and Komodo, both possessing an elongate filament that protrudes from the middle of the dorsal fin. It has also been observed and photographed on the north Sulawesi mainland near Lembeh Strait and at the Raja Ampat Islands, Papua Province. It was recently described by Allen and Kuiter (1999).

Table 4.12. Comparisons of fish faunas of the Togean and Banggai Islands.

Distribution	Number of Species
Banggai Islands	661
Togean Islands	596
Togean and Banggai Islands -shared	464
Banggai Islands only	146
Togean Islands only	100

Table 4.13. List of species that appear to be confined to either the Togean or Banggai Islands (note: most of these species may also occur in other regions within the Coral Triangle). Possible endemic species are indicated in bold print. Species indicated with an asterisk were also observed on the adjacent mainland peninsula.

Togean Islands	Banggai Islands
Cheilodipterus alleni	*Labracinus cyclophthalmus*
Neopomacentrus filamentosus	*Pseudochromis perspicillatus*
Pomacentrus cuneatus	*Apogon chrysopoma*
Cirrhilabrus aurantidorsalis	*Pterapogon kauderni*
Paracheilinus togeanensis	*Acanthochromis polyacantha*
Ecsenius sp. 1	*Chromis scotochilopterus* *
Meiacanthus sp.	*Chrysiptera bleekeri*
Ctenochaetus tominiensis	*Dischistodus prosopotaenia*
	Neoglyphidodon oxyodon
	Cirrhilabrus solorensis *
	Cirrhilabrus tonozukai *
	Halichoeres podostigma *
	Scarus schlegeli
	Meiacanthus vicinus

Paracheilinus togeanensis (Labridae)—A 48 mm SL male was speared in 25 m depth near the southern entrance to Batudaka Passage, Togean Islands (Site 23). It is unusual among species of *Paracheilinus* in lacking dorsal-fin filaments and having an elevated dorsal fin with a rounded posterior profile. The species is known only from the Togean Islands, where it was occasionally observed on rubble bottoms. It was recently described by Kuiter and Allen (1999).

Parachelinus cyaneus (Labridae)—Two specimens, 51–53 mm SL, were speared in 25 m depth at Bandang Island, near Banggai Island (Site 32). The species is similar to *Paracheilinus filamentosus* in shape, but has body markings reminiscent of *P. carpenteri* from the Philippines. It is known from northeastern Kalimantan and northern Sulawesi. It was recently described by Kuiter and Allen (1999).

Ecsenius sp. 1 (Blenniidae)—Five specimens, 20–25 mm SL, were collected with quinaldine (a fish anaesthetic) in 2 m depth near the southern entrance to Batudaka Passage, Togean Islands (Site 23). The species is related to *E. banda-*

nensis of southern Indonesia, and is characterized by a bright blue belly. It is thus far known only from the Togean Islands. It will be described by V.G. Springer and G.R. Allen.

Two additional fishes that were collected or observed during this survey exhibited unusual color patterns compared to equivalent populations from surrounding areas. *Pentapodus trivittatus* (Nemipteridae) from the Gulf of Tomini and Banggai Islands possesses a broad orange bar across the rear part of the body and adults also lack sharply defined dark stripes on the sides, which are characteristic of most western Pacific populations. *Pomacentrus burroughi* (Pomacentridae) is broadly distributed in the western Pacific and is generally plain dusky brown (including fins) with a faint yellowish spot on the basal portion of the soft dorsal fin. Individuals from the Togean and Banggai Islands differ in having an abruptly white caudal fin and lack the pale dorsal-fin spot.

Banggai Cardinalfish

The apogonid fish *Pterapogon kauderni* Koumans is apparently restricted to sheltered reefs and bays at or near the larger islands in the Banggai Group. It lives on shallow reefs (to a depth of 16 m) or among sea grass beds, usually in association with branching corals, *Millepora, Diadema* sea urchins, anemones, and the fungiid coral *Heliofungia actiniformis*. The male of this species, as in many other apogonids, incubates the eggs until hatching. However, its eggs are much larger (about 2.5 mm diameter) than those of other family members and the young sometimes shelter within the mouth cavity of the male parent, particularly when danger threatens. *Pterapogon kauderni* lacks a pelagic larval stage, unlike other apogonids and most reef fishes, for which dispersal is dependent on this stage of the life cycle. Hence, it has the most limited distribution of any of the estimated 250 species of cardinalfishes.

Although *P. kauderni* was first discovered in 1920 and scientifically described in 1933, it was all but forgotten until the author and Roger Steene collected and photographed it in 1994 (Allen and Steene, 1995; Allen, 1996). This beautiful fish also featured prominently in a lecture I gave at the Marine Aquarium Conference of North America at Louisville, Kentucky in September 1995. This exposure generated considerable interest among the aquarium-fish community and prompted an unsuccessful collecting expedition to the Banggai Islands by personnel from the Dallas Aquarium. However, by June 1996 it began to appear in large numbers in aquarium stores in America, Europe, and Japan.

The Marine RAP survey revealed that the Banggai Cardinalfish is common in certain habitat conditions; we found it at Sites 31, 32, 35, 43, and 44. However, it is being harvested for the pet trade at an alarming rate. Several villages we visited were engaged in the trade for this species, and Tampuniki Village on southern Peleng had approximately 5,000 fish in holding cages. These are eventually sold to dealers from Manado and Ujung Pandang for between 150–1,500 rupiahs per fish depending on the point of sale. This level of fishing pressure, combined with its extremely restricted distribution and low reproductive rate, seriously threatens the species' survival. It could well become extinct within the next decade if conservation measures are not initiated. It is therefore recommended that it be listed as a threatened species by IUCN and CITES.

Endemism

In view of the broad dispersal capabilities via the pelagic larval stage of most reef fishes, minimal endemism can be expected in the Togean-Banggai Islands. Prior to the RAP survey the Banggai Cardinalfish (*Pterapogon kauderni*) was the only endemic species reported from the region. Although additional surveys are required along the mainland portion of the Gulf of Tomini, there is a strong possibility that at least two other species (*Paracheilinus togeanensis* and *Ecsenius* sp. 1) may be endemic to the area. Both are presently known only from the Togean Islands. *Ecsenius*, in particular, is a genus which has many endemic species confined to relatively small areas.

Overview of the Indonesian fish fauna

Indonesia probably possesses the world's richest reef fish fauna. Stretching in an east-west direction for approximately 5,000 km and embracing 18,508 islands, the archipelago features a seemingly endless array of marine habitats. Although universally acclaimed as a leading reef fish locality, there is no comprehensive, published faunal list to substantiate this claim. It is largely based on various generic and family monographs and reports from various collecting expeditions. For example Allen (1991) listed 123 pomacentrid species (since updated to 152 species), the world's second highest total after Australia (which includes a number of temperate species). Allen et al. (1998) listed 87 species of angelfishes (Pomacanthidae) and butterflyfishes (Chaetodontidae) from Indonesia, which represent the highest world total for these combined groups. Springer (1988 and 1991) recorded 15 species of the blenniid genus *Ecsenius* from Indonesia, surpassing any other area. Moreover, the world's highest reef fish total (1,111 species) for a single restricted locality was recorded at Maumere Bay on the Indonesian island of Flores (Allen and Kuiter, unpublished). The CFDI regression formula yields a predicted total reef fish fauna for Indonesia of 2,032, which so far remains unsurpassed by any other country.

More than anyone else, Pieter Bleeker, a Dutch scientist who died well over a century ago, is responsible for our present knowledge of the Indonesian reef fish fauna. The importance of his voluminous ichthyological works in providing a solid foundation for our knowledge of Indonesian fishes cannot be understated. Considering that he was employed as an army surgeon during his tenure in Indonesia (1842–1860), the extent of his ichthyological activity was remarkable.

During a career that spanned 36 years Bleeker published 500 papers that include descriptions of an incredible number of new taxa: 406 genera and 3,324 species. Approximately 75% of these published articles were devoted to the Indonesian fauna. Bleeker's knowledge of Indonesian fishes, both freshwater and marine was outstanding. Revisions of various groups of Indo-Pacific fishes by modern researchers frequently attest to Bleeker's uncanny intuition and astute understanding of natural relationships.

Unfortunately, there have been few attempts since Bleeker's time, and particularly in recent years, to consolidate our knowledge of Indonesian reef fishes. The only exception is Kuiter's (1992) book, which includes underwater photographs of approximately 900–1000 Indonesian species.

The author is presently compiling a modern list of Indonesian reef fishes based on the results of the present RAP and other surveys that he conducted over the past three decades at Sumatra, Riau Islands, Java, Kalimantan, Bali, Komodo, Flores, Molucca Islands, and Papua Province. The study will also incorporate museum records and a literature survey. It will provide accurate figures to support Indonesia's claim of being the most diverse region on earth for coral reef fishes.

References

Allen, G.R. 1991. *Damselfishes of the World.* Aquarium Systems, Mentor, Ohio.

Allen, G.R. 1993. *Reef Fishes of New Guinea.* Christensen Research Institute, Publication No. 8. Madang, Papua New Guinea.

Allen, G.R. 1996. The king of the cardinalfishes. *Tropical Fish Hobbyist* 44(9): 32–45.

Allen, G.R. 1997. *Marine Fishes of Tropical Australia and South-East Asia.* Western Australian Museum, Perth.

Allen, G.R. 1998. Reef and shore fishes of Milne Bay Province, Papua New Guinea. In: T. B. Werner and G.R. Allen (eds.). *A rapid biodiversity assessment of the coral reefs of Milne Bay Province, Papua New Guinea.* RAP Working Papers 11, Conservation International, Washington, DC.

Allen, G.R. and M. Adrim. 2000. *Amblypomacentrus clarus,* a new species of damselfish (Pomacentridae) from the Banggai Islands, Indonesia. *Records of the Western Australian Museum* 20: 51–55.

Allen, G.R. and R.C. Kuiter. 1999. Descriptions of two new wrasses of the genus *Cirrhilabrus* (Labridae) from Indonesia. *Aqua, Journal of Ichthyology and Aquatic Biology* 3(4): 133–140.

Allen, G.R. and R.C. Steene. 1995. Notes on the ecology and behaviour of the Indonesian cardinalfish (Apogonidae) *Pterapogon kauderni* Koumans. *Revue Francaise d'Aquariologie* 22(1–2): 7–9.

Allen, G.R., R. Steene, and M. Allen. 1998. *A Guide to Angelfishes and Butterflyfishes.* Odyssey Publishing/Tropical Reef Research, Perth.

Kuiter, R.C. 1992. *Tropical Reef Fishes of the Western Pacific - Indonesia and Adjacent Waters.* Percetakan PT Gramedia Pustaka Utama, Jakarta.

Kuiter, R.C. and G.R. Allen. 1999. Descriptions of three new wrasses (Pisces: Labridae; *Paracheilinus*) from Indonesia and North-western Australia with evidence of possible hybridisation. *Aqua, Journal of Ichthyology and Aquatic Biology* 3(3): 119–132.

Myers, R.F. 1999. *Micronesian Reef Fishes* (Third Edition). Coral Graphics, Guam.

Randall, J.E., G.R. Allen, and R.C. Steene. 1990. *Fishes of the Great Barrier Reef and Coral Sea.* Crawford House Press, Bathurst, Australia.

Springer, V.G. 1988. The Indo-Pacific blenniid fish genus *Ecsenius. Smithsonian Contributions to Zoology* 465: 1–134.

Springer, V.G. 1991. *Ecsenius randalli,* a new species of blenniid fish from Indonesia, with notes on other species of *Ecsenius. Tropical Fish Hobbyist* 39(12): 100–113.

Chapter 5

Coral Reef Fish Stock Assessment in the Togean and Banggai Islands, Sulawesi, Indonesia

La Tanda

Summary

- Fishes were observed along a 100 m transect at various sites in the Togean and Banggai Islands, central Sulawesi.

- A total of 183 species belonging to 43 genera and 13 families were observed in the area.

- A total of 147 edible (commercial) species from 38 genera were classified as target fish. This segment of the fauna was dominated by *Caesio teres* and *Pterocaesio randalli* (family Caesionidae).

- The reef fish biomass in the area was estimated at 5.33-298.27 ton/km2.

- Thirty-six species of butterflyfishes, family Chaetodontidae, were observed. These fishes are considered indicators of healthy reef conditions. Local representation was dominated by *Chaetodon kleinii, C. trifasciatus,* and *Hemitaurichthys polylepis.*

Introduction

Increasing population and development has adversely affected terrestrial natural resources. There is increasing dependency by local communities, therefore, on the sea to supply basic food items. Coral reefs and their resident fish populations are an important commodity in this respect, and commercial harvesting of food and aquarium fishes is also a valuable source of income. However, basic scientific data is required for the sound management and utilization of these resources. This includes information on species diversity, distribution, abundance, and economic potential.

The present study was part of Conservation International's marine RAP survey of the Togean and Banggai Islands in cooperation with LIPI and Hasannudin University. The aim was to collect data for use in assessing the potential of marine resources, especially coral reef fishes. Hopefully the results can be used as a basis for the future management and sustainable use of the area's marine resources.

Methods

The study area was situated off the east coast of Central Sulawesi, and included the Togean Islands, a small portion of the Central Sulawesi Peninsula, and the Banggai Islands. Data were collected visually by scuba diving and recorded with pencil on an underwater slate. A visual census was employed, following the method of Dartnal and Jones (1986), with some modification to suit local conditions. Basically, a tape measure was used to delineate a 100 meter transect and observations were made for a distance of 5 m on either side of the tape, effectively forming a survey area of 1000m^2 per transect. The same transects used for the coral condition analysis were used at each site, giving a total of three transects per site at depths of approximately 3, 10, and 24 m. Fish were identified with the aid of various field guides including Kuiter (1993), Masuda and Allen (1987) and Allen (1997). Fish species were noted and the numbers were counted and recorded for each 100 m transect.

At each location, fish biomass was estimated by their number and average weight in tons/km^2. The calculation was based on the average length of each species and group size, using the method of Sparre and Venema (1992). Length was converted into weight according to the equation of W = aLb, where b = 3 and a = 0.05. Fish species were classified as either indicator or target species.

Indicator species

This group consists of species with a strong association with living coral reefs, and are assumed to be indicative of "healthy" reef conditions. For the purpose of this study the indicator species were solely members of the family Chaetodontidae, a group of brightly colored fishes popular with both divers and aquarists. Members of this family are all conspicuous diurnal residents of coral reefs, which are easily identified and counted due to their being solitary or living in small groups.

Target species

Target species are edible fishes, usually seen in local fish markets (ie. commercial fishes) that live on or near coral reefs. For the purpose of this study they are further subdivided as follows: Group A—solitary fishes; Group B—occur in large groups within the confines of the coral reef; Group C—schooling pelagic fishes often found in the vicinity of coral reefs. Quantitative data on solitary fishes or those that live in small groups were collected by actual count, whereas data for groups of schooling fishes such as those in the families Caesionidae and Acanthuridae were assigned to categories of abundance, as were some small pelagic fishes.

Results and Discussion

A total of 183 species of coral fishes in 43 genera, belonging to 13 families, were observed in the three areas surveyed (Appendix 6). At the 25 survey sites in the Togean Islands, 142 species were observed, belonging to 37 genera. A total of 68 species in 26 genera were found at three locations around the Central Peninsula, whereas 150 species from 37 genera were found at 19 sites in the Banggai Islands.

Target and indicator species occurring on coral reefs in the area belong to the following families (number of observed species in parentheses; see also Appendix 7): Chaetodontidae (36), Acanthuridae (26), Serranidae (26), Lutjanidae (18), Caesionidae (15), Siganidae (13), Carangidae (12), Lethrinidae (9), Nemipteridae (8), Mullidae (8), Haemulidae (6), Sphyraenidae (3) and Scombridae (2). The total includes 147 target species from 38 genera and 12 families, and 36 indicator species from 5 genera, all in the family Chaetodontidae.

The results indicate that there are generally more indicator and target species in the Togean and Banggai Islands compared to several locations previously surveyed in Indonesia by the author. These include Derawan Island (East Kalimantan), Tiga Islands (North Sulawesi), Watubela Islands (Central Mollucas), and Taka Bone Rate Islands (South Sulawesi). The numbers of species of target and indicator groups at each location are shown in Figure 5.1.

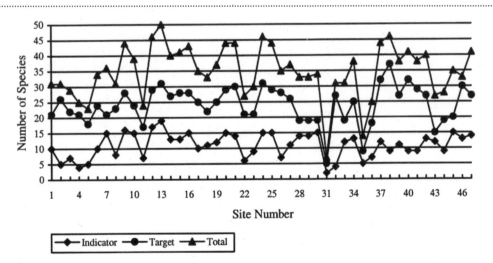

Figure 5.1. Total number of species of target and indicator groups at each site in the Togean and Banggai Islands.

Indicator species

A total of 36 indicator species belonging to five genera of Chaetodontidae were observed: 32 species in the Togean Islands, 19 species in the peninsular area, and 33 species in the Banggai Islands. The genus *Chaetodon* was represented by 27 species, *Heniochus* by five species, *Forcipiger* by two species, and *Hemitauricthys* and *Coradion* by a single species each (Appendix 6). The highest numbers of both individuals and species were recorded at Sites 12 and 13, located at Una-una Island in the Togean Group (Appendix 7).

The number of indicator species observed at the Togean and Banggai Islands exceeded the totals recorded by La Tanda (1996) at Biak, Papua Province (32 species), by Hukom (1994) at Bitung, North Sulawesi (25 species), and by Adrim and Yahmantoro, (1994) at the Tiga Islands, North Sulawesi (13 species).

The dominant species in terms of numbers of individuals in the Togean Group were *Hemitaurichtys polylepis* (22.9% of observed indicator species) and *Chaetodon punctatofasciatus* (9.2%). The dominant species in the peninsular area was *Chaetodon kleinii* (38.2%), and in the Banggai Islands *C. kleinii* (35.9%) and *H. polylepis* (10.4%) were the most abundant.

In terms of the percentage of occurrence at all surveyed sites, the most common species were as follows (Appendix 6): Togean Islands—*Chaetodon lunulatus* and *C. rafflesii* (80% of sites), *C. vagabundus* and *C. baronessa* (72%); Peninsular area—*C. kleinii, C. baronessa* and *C. trifascialis* (100%); Banggai Islands—*C. kleinii* (100%), *C. trifasciatus* (89.5%), and *Heniochus varius* (73.7%).

Chaetodon lunulatus and *C. kleinii* were commonly found in the Togean and Banggai Islands (Appendix 6). Both species are common throughout eastern Indonesia waters where there is a high degree of light penetration. Adrim and Yahmantoro (1994) and Adrim and Hutomo (1989) found that *C. kleinii* was the most common species in the Tiga islands (North Sulawesi) and in the Flores Sea (East Nusa Tenggara). La Tanda (1996) noted that *C. trifasciatus* (=*C. lunulatus*), *Forcipiger longirostris* and *C. vagabundus* were commonly found in clear waters of Biak, Irian Jaya. Conversely, *Chaetodon octofasciatus* was the most abundant chaetodontid in the Seribu Islands, Java Sea, in waters with relatively low light penetration (Adrim et. al., 1991). *Hemitaurichtys polylepis* was one of the most abundant butterflyfishes observed on outer reef dropoffs during the present survey. It frequently forms large aggregations. According to Allen, (1997) this species feeds predominantly on zooplankton.

Chaetodon burgesii was previously reported from Indonesian seas by Allen (1997), but is infrequently encountered. This rare species was observed at several locations in the Togean Islands, invariably adjacent to steep slopes near ledges and caves in depths greater than 20 m.

Target species

A total of 147 species in 38 genera and 12 families were recorded. Mainly adult fishes were noted, with juveniles being seen only occasionally. The general diversity of target species was relatively high in comparison with areas in Indonesia where similar studies have been conducted, for example Bitung, North Sulawesi and Derawan Island off East Kalimantan (Hukom, 1994).

Six of the target fish families contained mainly solitary reef dwellers and were assigned to Group A. This assemblage was dominated by lutjanids (34.5%). Groups B and C (schooling reef fishes and pelagic fishes respectively) each contained three families. Group B was dominated by caesionids (66.9%) and Group C by scombrids (67.4%).

The total target species for the various areas was as follows: Togean Islands—110 species in 32 genera and 12 families; Peninsular area—47 species in 22 genera and 10 families; Banggai Island—117 species in 32 genera and 11 families. Based on the number of individuals recorded during the transects the following species were the most abundant in Group A: Togean Islands—*Scolopsis margaritifer* (8.2%), *Lutjanus boutton* (8.2%), and *Lutjanus decussatus* (7.6%); Peninsular area - *Pentapodus caninus* and *Scolopsis bilinetaus* (9.5%); and Banggai Islands—*Pentapodus caninus* (17.8%) and *Parupeneus multifasciatus* (9.5%).

Dominant fishes in Group B were: Togean Islands—*Caesio teres* (33.0%) and *Pterocaesio randalli* (12.3%); Peninsular area—*Pterocaesio pisang* (43.9%) and *P. tile* (27.4%); and Banggai Islands—*Pterocaesio pisang* (10.9%) and *Caesio teres* (10.7%).

The most abundant species in Group C included: Togean Islands—*Sphyraena pinguis* and *Sphyraena jello* (31.4%); Peninsular area—*Caranx melampygus;* and Banggai Islands—*Rastrelliger kanagurta* (93.8%).

Based on their percentage of occurrence, the most commonly found target species were: *Lutjanus decussatus, Naso lituratus, Zebrasoma scopas* and *Ctenochaetus striatus* in the Togean Islands; *Scolopsis bilineatus, Parupeneus multifasciatus, Acanthurus lineatus* and *Acanthurus pyroferus* around the peninsular area; and *Parupeneus multifasciatus, Scolopsis margaritifer, Ctenochaetus striatus* and *Siganus vulpinus* in the Banggai Islands (Appendix 6).

The sites exhibiting a combination of the highest number of species and individuals of target (i.e., food) fishes included 12 and 14 (Una-una Island) in the Togeans, site 28 (Dongolalo Point) in the peninsular area, and Sites 32 (Bandang Island) and 47 (Makailu Island) in the Banggai Group. These sites were invariably characterised by good habitat diversity, including shallow reef flats and deeper slopes, as well as very clear water allowing for high levels of light penetration.

Fish biomass

The estimated lengths of each target and indicator species observed during the transects were converted into weight per unit area (ton/km²) as an indicator of overall biomass. Fish biomass for the survey sites ranged from 5.33–298.27 ton/km² (Table 5.1)

Biomass estimates for the three main areas were as follows: Togean Islands—10.50-238.73 ton/ km² (average 69.32 ton/km²); peninsular area—26.37-41.98 ton/km² average 32.18 ton/km²); and Banggai Islands—5.33-298.27 ton/ km² (average 67.43 ton/km²). The total average value of fish biomass in the three areas shows that the Togeans had the highest value (41.03%), followed by the Banggais (39.92%), and peninsular area (19.05%).

The biomass at a given site is obviously dependent on the size and number of individual fishes. The highest biomass in the Togean Islands was recorded at site 20 (Batudaka Island), and in the Banggai Islands at site 47 (Makailu Island). The relatively high biomass at Makailu was due to an abundance of schooling fusiliers (Caesionidae), including six species, of which *Pterocaesio tessellata, P. randalli,* and *Casio teres* were the most numerous. These fishes are considered good eating and consequently have high economic value.

Groupers (Serranidae) were the most speciose target family in the survey, being represented by 26 species. Survey observations indicated they are commonly captured by local fishermen and are highly valued. These fishes are generally sold fresh to the larger fishing companies but are also dried and salted. The abundance of groupers in the Togean-Banggai area was calculated at 0–12.67 individuals/1000 m², with the highest value recorded for site 12 on Una-Una Island (Table 5.2).

The average size of individual fish was relatively large, generally >20cm in 10–20 m depth. The number of individuals found in a particular area depends on the condition of the reef and the level of fishing activity. In areas where fishing was extensive, the number of individuals was relatively low. Una-una village, situated on the relatively remote Una-una Island, had the highest number of individuals for the eight species recorded from there: *Cephalpolis miniata, C. cyanostiqma, C. argus, C. leopardus, C. boenack, C. urodeta, Variola louti,* and *Plectropomus albimaculatus.* In addition, their average size was relatively large compared to other areas.

The chief factors affecting the abundance of coral fishes in the Banggai Islands included intensive use of traps (bubu), widespread hook-and-line fishing, and the common use of dynamite.

Table 5.1. Biomass of reef fishes from 47 sites in the Togean-Banggai Islands.
Legend: NC = non commercial fishes, COM = commercially important fishes.

Site No.	Biomass (ton/km²)			Number of families
	NC	COM	Total	
1	2.09	62.33	64.42	8
2	0.94	37.33	38.27	11
3	1.24	50.00	51.24	11
4	0.88	26.34	27.22	11
5	0.73	43.37	44.10	9
6	1.04	23.57	24.61	9
7	3.73	36.52	40.25	9
8	1.03	20.86	21.89	10
9	2.05	50.87	52.92	10
10	6.03	77.13	83.16	9
11	0.80	9.70	10.50	10
12	9.39	224.73	234.12	10
13	7.46	88.12	95.58	9
14	2.60	178.24	180.84	10
15	2.79	96.99	99.78	11
16	2.64	72.86	75.50	11
17	3.16	16.69	19.85	10
18	1.01	36.51	37.52	10
19	1.73	58.97	60.70	11
20	3.14	235.59	238.73	11
21	3.08	31.25	34.33	10
22	1.53	32.59	34.12	9
23	1.64	20.33	21.97	9
24	2.25	83.56	85.81	10
25	2.62	53.09	55.71	11
26	2.67	39.31	41.98	10
27	2.08	26.12	28.20	9
28	3.12	23.25	26.37	9
29	1.57	11.92	13.49	9
30	1.42	10.14	11.56	8
31	0.06	8.87	8.93	5
32	5.46	106.28	111.74	9
33	3.44	17.47	20.91	9
34	6.69	29.32	36.01	10
35	0.52	4.81	5.33	6
36	1.60	28.39	29.99	8
37	3.12	50.42	53.54	9
38	3.75	34.23	37.98	11
39	1.82	31.62	33.44	8
40	1.27	78.67	79.94	12
41	3.37	187.13	190.50	10
42	2.83	91.72	94.55	9
43	2.12	18.49	20.61	8
44	0.93	24.68	25.61	8
45	3.71	20.38	24.09	10
46	2.41	60.38	62.79	10
47	10.63	409.48	420.11	10

Table 5.2. Density of grouper for the 47 sites surveyed in Togean-Banggai Islands in October-November 1998. Values are expressed as the number of individuals per 1000 m².

Site No.	Density (ind./100 m²)	Site No.	Density (ind./100 m²)
1	5.57	25	6.00
2	4.67	26	5.50
3	0.67	27	6.33
4	1.67	28	0.00
5	1.33	29	0.33
6	0.33	30	0.67
7	5.67	31	0.50
8	0.50	32	1.67
9	2.33	33	7.00
10	2.00	34	2.33
11	5.00	35	0.00
12	12.67	36	2.67
13	5.33	37	1.33
14	7.67	38	1.00
15	3.33	39	1.00
16	0.67	40	1.67
17	6.33	41	2.00
18	1.00	42	3.00
19	2.33	43	1.00
20	0.67	44	3.00
21	1.33	45	0.33
22	1.00	46	0.33
23	0.67	47	4.00
24	2.00		

Conclusion

The seas surrounding the Togean and Banggai Islands support a highly diverse fish fauna with relatively high numbers of individuals. The area therefore has good potential for sustained reef fisheries as well as marine-based tourism with an emphasis on fishing and diving activities such as underwater photography.

References

Adrim, M. and M. Hutomo. 1989. Species composition, distribution and abundance of Chaetodontidae along reef transects in the Flores Sea. *Netherlands Journal of Sea Research* 23 (2); 85–93.

Adrim, M., M. Hutomo, and S.R. Suharti. 1991. Chaetodontid fish community structure and its relation to reef degradation at the Seribu Islands reefs, Indonesia. *In:*

A.C. Alcala (ed.). *Proceedings of the Regional Symposium on Living Resources in Coastal Areas*. Manila, Philippines. pp. 163–174.

Adrim, M. and Yahmantoro. 1994. Studi pendahuluan terhadapfauna ikan karang di perairan P.P. Tiga, Sulawesi Utara. Wisata Bahari Pulau-Pulau Tiga (Tundonia, Tenga, Paniki) Sulawesi Utara. Lembaga Ilmu Pengetahuan Indonesi, Puslitbang Oseanologi. Proyek Penelitian dan Pengembangan Sumberdaya Laut. Jakarta. pp. 29–44.

Allen, G. 1997. *Marine Fishes of Tropical Australia and Southeast Asia*. Western Australian Museum.

Dartnall, H.J. and M. Jones. 1986. *A Manual of Survey Methods: Living Resources in Coastal Areas*. ASEAN—Australia Cooperative Programme on Marine Science Handbook. Townsville: Australian Institute of Marine Science.

Hukom, F.D. 1994. Keanekaragaman jenis ikan karang di daerah Bitung dan Sekitarnya. Laporan Kemajuan Triwulan IV. Tahun Anggaran 1994/1995.

Kuiter, R.H. 1992. *Tropical Reef-Fishes of the Western Pacific, Indonesia and Adjacent Waters*. Gramedia, Jakarta.

La Tanda. 1996. Komunitas ikan kepe-kepe di daerah terumbu karang perairan Biak, Irian Jaya. Perairan Maluku dan Sekitarnya. Balai Penelitian dan Pengembangan Sumberdaya Laut, P3O-LIPI Ambon. Vol. 11. pp. 79–88.

Masuda, H. and G.R. Allen. 1987. *Sea Fishes of the World (Indo Pacific Region)*. Yama-Kei, Tokyo.

Sparre, P. and S. Venema. 1992. *Introduction to tropical fish stock assessment, Part 1*. FAO Fisheries Technical Paper 306. FAO, Rome.

Chapter 6

Exploitation of Marine Resources in the Togean and Banggai Islands, Sulawesi, Indonesia

Purbasari Surjadi and K. Anwar

Summary

- Nearly 80% of the inhabitants of the Togean and Banggai Islands live in coastal areas; most earn their living from the sea.

- Fishing methods commonly employed include poison (natural and cyanide), basket traps (bubu), dynamite, nets, hook-and-line, and spear fishing.

- Most of the catch is for food, and is eaten fresh or preserved (usually dried and salted) for future consumption.

- The recent Indonesian monetary crisis appears to be at least partly responsible for the increased use of dynamite over the past two years. Economic pressure frequently forces fishermen to abandon traditional methods in favor of more destructive techniques.

- A few species of ornamental fishes, as well as species such as *kerapu* and *sunu* (types of groupers, family Serranidae), *mamin* or napoleon wrasse (Labridae), are specifically targeted for live capture. This type of fishing is becoming more popular as the demand for live fishes increases both for the Asian restaurant trade and ornamental fish industry.

- One of the biggest problems is a lack of enforcement of conservation laws. Fishing of protected species, such as sea turtles, and illegal fishing methods, particularly the use of dynamite and potassium cyanide, are common in both the Togean and the Banggai Islands.

- The Togean Islands have good potential for the development of marine ecotourism.

Introduction

The Togean/Banggai Islands are rich in marine natural resources, and it is important to ensure their conservation and long-term economic sustainability. About 80% of the inhabitants are involved in fisheries, and the entire population depends on marine natural resources in one way or another. A variety of fishing techniques are employed, as well as special methods for capturing molluscs and turtles, which are used for both food and ornamental decoration. Unfortunately, fishing pressure is heavy, and overfishing is an obvious consequence. Over-exploitation has caused some mollusc species to decline in numbers, such as giant clams (*Tridacna*) and lola (*Trochus*). In addition, turtles are hunted despite their protection by law as endangered species.

Over-utilization of marine products poses a serious threat to the populations of both island groups. To protect and carefully manage the existing natural resources, there is an urgent need for accurate information on resource availability and utilization. Therefore, one of the goals of the present RAP survey was to gather baseline information for future initiatives related to conservation and sustainable exploitation of marine resources.

Methods

The main survey methods consisted of direct observations of fishing operations and their socio-economic effects through interviews with village residents. Residents were also questioned at length about their perception of the importance of marine resources and the need for conservation.

Results and Discussion

Fishing techniques

Nearly 80% of the inhabitants of the Togean and Banggai Islands live in coastal areas, and most of them earn their living from the sea. Local residents use several fishing techniques that range from simple, environmentally friendly ones to more elaborate and harmful methods. The appropriate technique depends largely on the type of fish that is targeted and the way it is utilized (e.g., for immediate consumption or maintained alive). The various techniques are described in the following paragraphs.

Poison (traditional poison/tuba and potassium cyanide)

The use of poison is a traditional fishing method. The substance used by local fishers is similar to the commercial product known as rotenone that is used by ichthyologists to collect scientific samples. It is derived from the root of derris plants, which either grow wild or are cultivated. The roots are crushed and the resultant milky liquid is dispersed in the sea, often in "closed" situations, such as a large rocky pool at low tide.

This substance causes constriction of the gill capillaries, thus preventing normal respiration. It is a comparatively harmless method in that it works only in small, confined areas (i.e., a small section of a reef flat at low tide), is specific towards fishes, and has minimal environmental impact. Prolonged exposure usually kills the fish, but with only a mild dose, the fish can be revived. A more modern technique involves the use of potassium cyanide, which not only kills or stuns fishes, but also corals and other invertebrates. Divers generally dispense the potassium cyanide using plastic squeeze bottles. It is squirted at fishes hiding in reef crevices. No special diving equipment is necessary, although the use of hookah was observed on one occasion near Banggai Island. One of the results of cyanide use is coral bleaching, and this type of damage is difficult to differentiate from "natural" bleaching due to rise in sea temperatures.

Basket traps (bubu)

The *bubu* is a traditional basket trap commonly made from bamboo, rattan, or fine wire mesh in the shape of a cylinder or rectangular box. The traps are variable in size, but may measure as large as 1.5 (1.0 (0.5 m. Each is equipped with an inverted cone-shaped opening that allows an easy entrance, but once the fish enters it is nearly impossible for it to find the opening again. The traps are placed on the bottom by free (breath-hold) diving or with the aid of SCUBA. They are secured to the sea floor by metal anchors, or coral rocks are often piled on top, which also help to conceal the traps. As many as 5-10 *bubu* may be placed in a relatively small section of the reef and left overnight. This method is used for catching fishes alive, an important advantage consid-

ering the lack of cold storage facilities and the lucrative trade in live fishes (see discussion below). Extensive use of *bubu* causes significant reef damage due to corals being broken for use in concealing the traps.

Dynamite fishing

Dynamite fishing is common in the Togean and Banggai Islands despite its illegal status. We heard at least five separate blasts during our survey activities and found extensive blast damage at several sites. Fishermen generally make their own explosives by stuffing fertilizer and match heads into a bottle. A more sophisticated method consists of using a simple detonator attached to a battery with a wire. Dynamite is often used to catch schooling fishes or concentrations of larger reef fishes.

Nets

A variety of nets are used depending on the target species. Vertical nets set across the migration paths of certain pelagic species are commonly employed, and floating circular nets are also used in Walea and in a few places in the Banggai Islands. Coral netting is an additional method in which sections of reef are covered with mesh and sheltering fishes are then prodded from their hiding places into the net. This technique is effective for grouper and other large reef fishes that feature in the live fish trade.

Rumpong is a common method that consists of vertical nets stretched across four poles, which are hammered into the sea floor, thus forming a rectangular-shaped enclosure. It is illuminated during the night by suspended lights, often from a small fishing boat, which serves to attract small baitfish and larger pelagic fishes such as tuna. The method is very effective and frequently results in large catches. For example, it is estimated that between 0.5–2.0 tons of fish are caught per night by a single rumpong vessel. The catch is equally divided between the rumpong owner and the fishermen. The lucrative financial reward of this fishery is responsible for its rapid expansion in recent years. In the Walea area alone there are approximately 200 rumpongs in operation, each using an average of two nets per operation.

Hook-and-line

Fishing lines with an attached hook are commonly used in the area. Fishing lines are rigged in a variety of ways, and there is comprehensive local terminology for describing them. The traditional terms include: *renjo, tondak, barita, batapel,* and *buang batu.*

Spear fishing

Spears are used in still, clear water to kill slow-moving fish. They are thrown directly from small fishing craft from a standing position. Fishermen also shoot fish with spears while free diving.

Catch Utilization

Most of the catch is intended for food and is consumed while fresh or preserved (usually dried and salted) for future consumption. A summary of the fishes and other marine organisms consumed by local villagers is given in Appendix 8.

Live Fish Trade

Several fishes, such as *kerapu* and *sunu* (types of groupers, family Serranidae), *mamin* or napoleon wrasse (Labridae), and a few ornamental fishes, are specifically targeted for live capture. Fishing methods include hook-and-line, coral netting, basket traps and poison (in low doses). This type of fishing is becoming more popular as the demand for live fish increases, both for the Asian restaurant trade and ornamental fish industry.

Fishermen are paid for live fishes according to species, size, and condition. *Cephalopholis miniata* (super kerapu/grouper), the most expensive grouper, sells for between 45,000–90,000 Rp. per kg. The most sought after-fish, *Cheilinus undulatus* (mamin/napoleon wrasse) ranges between 70,000–100,000 Rp. per kg when alive but dead individuals bring only 2,000–4,000 Rp. per kg. Prices of damaged or injured fishes are also drastically reduced, particularly if there is little chance they will survive shipping.

About 20 storage-tank stations for live fishes are located in the Togeans, at small islands such as Tongkabo, Salakan, Pautu, Milok, Kabalutan, Anam, Kuling, Kinari, Bomba and a few other places in the Walea Group. Storage-tank stations are situated in the Banggai Islands at Banggai Island (in Banggai Bay), Peleng Pasibata, Nal, and at one village (Tugong Sagu) on Silumba Island. These holding facilities are usually owned by merchants from Ujung Pandang, Jakarta or Manado. There are also many smaller storage-tank stations directly owned by the fishermen. These are used to collect fishes before they are transferred to larger holding facilities. The catch is ultimately exported to Hong Kong, Singapore, China and Thailand through Manado every three to six months.

Although common in the Banggai Group, the collection of live ornamental fishes was not observed in the Togean Islands. Relatively few species are involved. The two most popular ornamentals are the Banggai Cardinalfish (*Pterapogon kauderni*) and the Blue-ringed Angelfish (*Pomacanthus annularis*).

As a result of its rediscovery in 1994, the Banggai Cardinalfish has surged to prominence on the world aquarium market. When first exported in early 1996 this fish created a sensation due to its beauty and unusual biology (males are oral egg incubators). Initially specimens retailed for more than US $100 per fish. Due to the large volume of fish being exported the price fell to about $30–50 per fish by late 1998. Banggai fishermen commonly fetch a price of 60 to 100 Rp ($1 US = 9,395 Rp. in 1998) per fish. Near Banggai Harbor, cardinalfish are sold for between 2,500–3,000 Rp. per kg (30 to 40 fish usually weigh one kilogram). Merchants in Manado, in turn, command 1,200–2,000 Rp. per fish. Cardinalfish are easily caught with fine-mesh nets while free diving or wading at low tide. They are then stored in netted enclosures until shipped to exporters in Manado.

Shellfish

Molluscs are more commonly used as food in the Banggai Islands than in the Togeans, judging from the large number of discarded shells seen around villages. Virtually all edible species are gathered for food by the Banggai Islanders and are consumed either fresh or preserved (dried and salted). In contrast, the Togean people eat only large molluscs such as *kima/* giant clam (Tridacnidae), although molluscs with ornate shells are also collected and used as decoration, both there and in the Banggai Islands.

Pearl-oysters are by far the most valuable commercial molluscs collected by local fishers. Pearls from natural oysters are considered more precious than artificially cultured ones, and are therefore in great demand. Unfortunately, this situation has caused over-fishing of the natural population, although in a few places such as the relatively remote Una-una Island, pearl oysters are still plentiful. Fishermen generally earn 25,000–40,000 Rp. per pearl.

Conservation Problems

One of the biggest problems is a lack of enforcement of conservation laws. Fishing of protected species, such as sea turtles, and illegal fishing methods, particularly the use of dynamite and potassium cyanide, are common in both the Togean and the Banggai Islands. Law enforcement is extremely difficult due to a shortage of staff and resources (boats, weapons, etc.). The total disregard of conservation laws and consequent over-harvesting poses a serious threat for the long-term sustainability of marine resources.

The recent Indonesian monetary crisis appears to be at least partly responsible for the increased use of dynamite over the past two years. The reason is that fishers have been forced to increase catches in order to keep pace with inflated prices for essential commodities (e.g., rice) and the devaluation of the Indonesian rupiah. The resulting effect is over-harvesting of resources and soaring prices for fish products. For example the price of salted fish ranged from 2,500 to 3,000 Rp. per kg in April 1998 compared to the November 1998 level of 6,000 to 9,000 Rp. per kg.

Economic pressure frequently forces fishermen to abandon traditional methods in favor of destructive techniques such as blasting. Sadly, they do not fully comprehend the dangers associated with these practices, particularly the long-term effects to the environment and non-sustainability of this type of fishing. Many local fishers that were interviewed suggested that dynamite is mainly used by outsiders from Pagimana and other parts of adjacent mainland Sulawesi. However, during the survey we noted that dynamite was being used by villagers from Kabalutan, Milok, Pautu and Papan. In defence of this method, the local fishers claimed that traditional methods were no longer economical and they were forced to use dynamite in order to survive. Despite an abundance of natural resources, most people living in the Banggai and Togean Islands are poor. For example, 29 of the 37 villages (*desa*) in the Togean Islands are part of the government special project for underdeveloped villages. A similar situation exists in the Banggai Islands, judging from our observations of housing and general living conditions.

There is considerable doubt as to whether the present rumpong harvest around Walea can be sustained indefinitely. It is conservatively estimated that 200 rumpongs are in operation five times per year resulting in an average annual production of pelagic fishes of 20,000 tons. Moreover, many rumpong are positioned in close proximity, which causes serious social conflicts among the various owners. In addition the nets are positioned haphazardly, creating a danger for marine navigation. There appears to be a compelling case for government intervention to control the number of rumpong and to institute guidelines for their operation, particularly in the Walea Regency.

Terrestrial Problems

Forest destruction has mainly ceased in the Togean Islands, but occasional logging and agricultural land clearing does occur. During the RAP survey of the two island groups, we saw very little evidence of primary forest. Coconut and cocoa plantations occupying former forest areas were commonly seen. There is a need for more effective management of terrestrial wildlife resources. It is reported that the Togean

macaque (*Macaca togeanus*), hornbill (*Ceros cassidix*), coconut crab (*Birgus latro*), parrots (Psittacidae) and saltwater crocodile (*Crocodylus porosus*) are often hunted.

State of Marine Ecotourism

The Togean Islands, situated in Tomini Bay, are generally more sheltered from prevailing weather than the Banggai Group, and marine ecotourism is better developed there. The islands possess a wide assortment of marine environments including fringing reefs, barrier reefs, patch reefs, and atolls. Major attractions include the outer barrier reef off Malenge Island and outstanding atoll conditions at Pasir Tengah and Pasir Batang. The Togeans are also characterized by high terrestrial biodiversity, which further enhances its attractiveness as an ecotourist destination.

The Togean Islands are strategically located on the tourist route connecting north and south Sulawesi. In the past three years, tourism has significantly grown in the area, reflected by more and better facilities for transportation and accommodation. This includes construction of 16 inns and cottages providing 116 rooms. Approximately 1500–1800 tourists visited the Togeans in 1995, which increased to about 2300 in 1996 and 2800 in 1997 (data provided by the administration of Una-Una subdistrict).

By contrast, ecotourism is poorly developed in the Banggai Islands, although the area does have potential. A number of good diving areas were identified during the present survey, including a "world class" site (47) at Makailu Island. But in general, tourist accommodation is either lacking or of poor quality. Nevertheless, the Banggai Island Tourism Department informed us that the area is being promoted and was recently visited by the Minister of the Environment. However, because of the low volume of tourist traffic, facilities often fall into disrepair. Lack of investment in tourist facilities and their promotion are two of the main reasons for the scarcity of tourists.

Appendicies

Appendix 1

Coral species recorded at the Togean and Banggai Islands, Sulawesi, Indonesia

Species	Site Records
Family Astrocoeniidae	
Stylocoeniella armata	15, 19, 22, 25, 33
Stylocoeniella guentheri	4, 9, 15, 20, 21, 32, 39, 41, 47
Family Pocilloporidae	
Madracis kirbyi	12, 16, 19
Pocillopora ankeli	13
Pocillopora damicornis	2, 3, 6, 7, 14, 16, 17, 19–23, 29, 30, 32, 33, 36, 41, 44, 47
Pocillopora eydouxi	7, 9, 13, 14, 16, 24, 25, 30, 32, 33, 36, 39, 40, 45–47
Pocillopora meandrina	1, 9, 13, 14, 24, 25, 29, 30, 32–34, 36, 38–40, 46, 47
Pocillopora verrucosa	1–3, 5–7, 10, 12–16, 18–25, 31–33, 35–38, 40–47
Seriatopora caliendrum	1–3, 5, 6, 8, 17, 20, 22, 25, 32–34, 37, 39, 41, 43, 45, 47
Seriatopora dendritica	36
Seriatopora hystrix	1, 2, 4, 5, 7–19, 21, 22, 24, 25, 29, 32–37, 42, 44, 47
Stylophora pistillata	1–7, 9, 10, 12, 14, 15, 17–23, 25, 29–34, 36–40, 42, 44–46
Palauastrea ramosa	2, 3, 18, 39, 42, 44, 45
Family Acroporidae	
Montipora aequituberculata	1, 7, 15, 21, 31, 45
Montipora cactus	7, 13, 15, 21, 24, 25, 31, 44, 45
Montipora caliculata	34, 38, 40
Montipora sp. yes	25, 32, 36
Montipora capitata	2–5, 14, 17, 19–24, 42
Montipora confusa	7, 10, 25, 30–34, 36–38, 41, 44, 45
Montipora corbettensis	16, 46
Montipora crassituberculata	16, 20, 24, 25
Montipora danae	13, 22
Montipora delicatula	14, 32
Montipora florida	2, 4, 11, 17, 22, 23, 31
Montipora foliosa	7, 12, 17, 24, 25, 30, 31, 33, 42, 45

Species	Site Records
Montipora foveolata	25, 46
Montipora hirsuta	3, 7, 13, 22
Montipora hispida	7, 13, 15, 16, 18, 20, 21, 24, 25, 29, 30, 32, 35, 36, 41, 42, 44–47
Montipora informis	10, 29, 30, 32, 33, 43
Montipora mollis	32, 41, 47
Montipora monasteriata	17
Montipora palawanensis	42
Montipora samarensis	33
Montipora tuberculosa	3, 4, 10, 24, 25, 32, 34, 38, 43, 44, 46, 47
Montipora venosa	5, 11
Montipora verrucosa	3, 10–14, 17, 18, 25, 31, 32, 37, 38, 42
Montipora verruculosus	41
Anacropora forbesi	2, 20, 22, 42
Anacropora matthai	1, 23–25, 29, 34, 35, 42
Anacropora puertogalerae	31, 42, 43, 45
Anacropora reticulata	4, 11, 18, 20, 22, 25, 29, 41, 42
Anacropora spinosa	2, 11, 35, 42
Acropora abrolhosensis	42
Acropora aculeus	2, 3, 12, 18–24, 31, 42, 44, 46
Acropora acuminata	30, 32–34, 40, 41
Acropora aspera	13
Acropora austera	38, 45, 46
Acropora batunai	7, 9, 12, 17, 19
Acropora brueggemanni	1, 12, 14, 15, 25, 29, 31, 34, 42, 45, 46
Acropora clathrata	1, 11, 12, 29, 33
Acropora cylindrica	8, 10, 11, 15, 17, 19–24
Acropora cytherea	1, 5, 10, 24, 30, 36, 43
Acorpora derawanensis	2, 3, 42, 43
Acropora divaricata	14, 16, 30, 32–34, 36, 38, 41, 44, 47
Acropora echinata	3–5, 9, 11, 18, 19, 35, 45
Acropora elegans = magnifi	2, 9, 12, 16, 23
Acropora fastigata	4, 5, 10, 16–20, 22–24
Acropora filiformis	23
Acropora florida	1, 3, 6, 13, 14, 22, 24, 25, 29, 30, 32–34, 42–47
Acropora formosa	1, 7, 9, 16, 17, 22, 24, 30, 33, 40, 42, 45, 47
Acropora gemmifera	1, 6, 8, 9, 12, 20, 37, 40, 41, 43, 45–47
Acropora grandis	43, 45
Acropora granulosa	1–3, 7, 9, 10, 12, 14, 16, 17, 19, 20, 25, 31, 32, 38, 41, 46, 47
Acropora horrida	29, 31, 32, 42, 43, 45
Acropora humilis	2, 5, 6, 16, 21, 24, 31–34, 36–39, 41–47
Acropora hyacinthus	1, 5, 7, 10, 13, 14, 16, 19, 21, 24, 25, 32, 37, 40, 45, 46
Acropora indonesia	8, 16, 21, 24, 25, 33, 34, 40, 41, 46, 47
Acropora jacquilinae	3, 15–17, 19–21, 23, 34
Acropora latistella	13, 24, 25, 32, 35, 36, 47
Acropora longicyathus	1, 24, 29, 42, 44, 45, 47
Acropora loripes	1, 3, 4, 6–12, 14, 16–22, 25, 33, 34, 37, 38, 40–42, 46, 47
Acropora microphthalma	45

Species	Site Records
Acropora millepora	1, 5, 6, 9, 13, 16, 21, 22, 24, 25, 29, 32–34, 36, 38, 40, 41, 44–47
Acropora monticulosa	13, 30, 40, 45
Acropora nasuta	6, 13, 21, 22, 45
Acropora nobilis	11, 30, 32, 33, 36
Acropora palifera	1–7, 9–12, 15–22, 24, 25, 30, 33, 34, 36–38, 40–47
Acropora pulchra	4, 5, 11, 18, 22, 45
Acropora robusta	9, 24, 25, 36, 40, 45, 46
Acropora samoensis	3, 6, 30
Acropora secale	1, 25, 38
Acropora selago	1–5, 7, 10, 16–19, 21–23, 31, 38, 43, 45–47
Acropora sesekiensis	31
Acropora simplex	21, 23
Acropora tenella	
Acropora tenuis	6, 8, 9, 13, 16, 21–24, 29, 30, 32, 36, 38, 40, 41, 45, 46
Acropora togianensis	3–6, 8, 11, 18, 22–24
Acropora turaki	18, 22
Acropora valenciennesi	10, 14, 19, 24, 29, 30, 32, 40, 41, 43
Acropora valida	45, 46
Acropora vaughani	7, 12, 24, 29–34, 36, 45, 46
Acropora willisae	33
Acropora yongei	9, 13, 17, 24, 25, 30, 32, 33, 44, 47
Astreopora expansa	24
Astreopora gracilis	8–10, 19, 21, 25, 46
Astreopora randalli	14, 16–18, 20, 24, 37, 38, 43
Astreopora listeri	5, 22, 38
Astreopora myriophthalma	7, 8, 11, 13, 16, 24, 34, 36, 38, 40, 46
Astreopora suggesta	20, 24

Family Poritidae

Species	Site Records
Porites annae	5, 7–9, 14, 15, 20–23, 25, 41, 42
Porites attenuata	3, 4, 8, 16, 18–20, 22
Porites cumulatus	23
Porites cylindrica	2–6, 8, 10, 11, 16–24, 29–34, 36, 38–43, 45, 47
Porites evermanni	6, 9, 13, 16, 44
Porites horizontalata	1, 6–10, 14–16, 18, 20, 22, 24, 39
Porites monticulosa	2, 3, 5–9, 12–14, 16, 17, 19–22, 24, 41–43, 47
Porites negrosensis	2, 3, 5, 11, 19, 20, 22, 23, 31, 44
Porites nigrescens	2, 3, 6, 8
Porites rugosa	17, 19, 21
Porites rus	1–4, 6, 8–10, 12, 14–17, 19–22, 24, 25, 31, 36, 39, 42–44, 46, 47
Porites vaughani	8, 9
Goniopora columna?	10, 11, 16, 21, 30, 31, 35
Goniopora pendulus	31, 42
Goniopora stokesi	33
Goniopora tenuidens	40
Alveopora allingi	4
Alveopora catalai	1, 29, 31, 35

Species	Site Records
Alveopora excelsa	31
Alveopora verrilliana	42

Family Siderasteridae

Pseudosiderastrea tayami	3
Psammocora contigua	2, 8, 14, 17, 19, 21, 23, 31
Psammocora digitata	5, 9, 10, 14, 16, 22, 30, 31, 33, 39, 41, 42
Psammocora haimeana	41
Psammocora nierstraszi	6, 9, 13–15, 19, 24, 30, 31, 39, 41, 43, 46, 47
Psammocora profundacella	1, 4, 12–14, 16, 19, 20, 25, 32, 36, 42
Psammocora superficialis	3, 21, 31, 33
Coscinaraea columna	30, 33

Family Agariciidae

Pavona bipartita	1, 8, 17, 25, 31, 34, 41, 44
Pavona cactus	2–5, 7–12, 14, 15, 17–24, 35, 40, 42, 44, 45
Pavona clavus	1, 2, 12, 17, 19–22, 25, 29, 30, 32, 40, 42, 43
Pavona decussata	3–5, 9, 10, 16, 29, 35, 40, 41, 43, 45, 46
Pavona explanulata	1, 4, 6–9, 13–17, 20, 25, 31, 37, 38, 40, 41, 43–47
Pavona duerdeni (minuta)	1, 12, 25, 30, 32–34, 36, 41, 47
Pavona varians	1, 3, 4, 6, 9, 10, 12–17, 19–22, 24, 25, 29, 32, 38, 40, 41, 45–47
Pavona venosa	9, 13, 14, 44, 45
Pavona minuta (=xarifae)	1, 13, 15, 21, 25, 47
Leptoseris amitoriensis	29, 47
Leptoseris explanata	2, 7–9, 12–14, 19, 23–25, 32, 37, 47
Leptoseris foliosa	7
Leptoseris gardineri	5, 17, 19, 24, 35
Leptoseris hawaiiensis	2, 7, 9, 12–17, 19–21, 24, 25, 38, 47
Leptoseris mycetoseroides	7, 12, 14–16, 21, 24, 25, 41, 44
Leptoseris papyracea	1, 5, 19, 29
Leptoseris scabra	1–5, 7, 11–15, 19, 20, 24, 25, 38, 44, 47
Leptoseris striata	2, 15, 47
Leptoseris yabei	1, 10, 12, 16, 20, 29, 31, 38, 43, 46
Gardineroseris planulata	1, 3, 5–7, 9, 11–16, 23, 29, 31, 33, 36, 39, 41, 43, 44, 46, 47
Coeloseris mayeri	3, 5–11, 13, 15–18, 20–25, 29, 30, 40–46
Pachyseris foliosa	2–5, 8, 25, 29, 35, 36, 43–45
Pachyseris gemmae	1, 2, 5, 7–9, 12, 14, 15, 17, 19, 20, 37, 41, 42, 44–46
Pachyseris involuta	8, 11, 17–24, 31
Pachyseris rugosa	2–4, 6–9, 11, 13–17, 19, 21, 23, 24, 30–33, 41–46
Pachyseris speciosa	1, 3–5, 7–12, 14–16, 18–21, 23, 24, 29, 31, 32, 35, 37, 38, 40–42, 45–47

Family Fungiidae

Cycloseris costulata	18, 44
Cycloseris tenuis	4, 39
Cycloseris vaughani	42
Diaseris distorta	29, 34
Diaseris fragilis	34

Species	Site Records
Fungia concinna	12
Fungia fralinae	3
Fungia fungites	4, 5, 7, 9, 14, 17, 19, 21, 24, 29, 30, 32, 36–40, 42, 45, 46
Fungia granulosa	1, 7, 36–39
Fungia horrida	1, 5–7, 9, 10, 12, 14–22, 24, 25, 32–34, 37, 38, 40, 44–47
Fungia klunzingeri	4, 8, 11, 17–19, 21, 23, 24, 37, 42, 45
Fungia moluccensis	10, 11, 18, 20, 29, 37, 38
Fungia paumotensis	2, 4, 6, 11–13, 15, 16, 18, 20–22, 24, 31, 33, 35–41
Fungia repanda	3, 9, 13–15, 17, 44
Fungia scruposa	4, 9, 12, 13, 16–18, 22, 23, 25, 31, 33, 34, 37, 42–44, 46
Fungia scutaria	13, 14, 34, 46
Heliofungia actiniformis	2, 3, 5, 10, 16, 18, 20–24, 29, 31, 35, 39–42, 44, 45
Ctenactis albitentaculata	1, 16, 17, 19–21, 30, 34, 39, 40, 43, 44
Ctenactis crassa	2, 3, 5–9, 13, 19–22, 24, 33, 37, 39, 42, 45, 47
Ctenactis echinata	1, 5, 6, 8, 9, 12, 14–17, 19, 21, 22, 24, 25, 30–34, 36, 38, 39, 41, 43–47
Herpolitha limax	2, 3, 5–7, 9–13, 16–23, 24, 25, 29–34, 37, 40–42, 45
Polyphyllia talpina	2, 5, 7, 8, 10, 16, 20–22, 32–34, 38, 39, 41–44, 46
Halomitra clavator	19, 21
Halomitra meiere	34
Halomitra pileus	1, 7, 9, 12, 14–16, 24, 25, 29, 31–42, 44–47
Sandalolitha dentata	3, 9, 13, 19, 20, 24, 38
Sandalolitha robusta	2–6, 8–12, 16, 18, 19, 21, 22, 24, 32–34, 37–42, 44–47
Zoopilus echinatus	1, 5, 19, 23, 25, 31, 35, 37, 45
Lithophyllon undulatum	18, 44
Podabacia crustacea	4, 8, 13, 17, 18, 22, 24, 25, 32, 38, 39, 41, 44, 45, 47
Podabacia motuporensis	1–4, 6, 7, 9, 10, 12, 14–17, 21, 38, 42

Family Oculinidae

Galaxea astreata	1, 6, 8, 14, 16, 24, 31, 36, 40–43
Galaxea cryptoramosa	30, 32, 34
Galaxea fascicularis	1–3, 9, 11–25, 29–34, 36–40, 42, 44–47
Galaxea horrescens	1–12, 16–24, 29, 31, 35, 42–45
Galaxea longisepta	1, 4, 7, 12–15, 20, 21, 24, 25, 35
Galaxea paucisepta	40–42, 44, 46

Family Pectinidae

Echinophyllia aspera	25, 31–34, 37, 38, 46, 47
Echinophyllia costata	35, 43
Echinophyllia echinata	3–5, 11–13, 19–22, 25, 38, 39, 47
Echinophyllia echinoporoides	3, 43
Echinophyllia orpheensis	4, 5, 22
Echinophyllia patula	25, 30, 32, 33, 37, 40, 46
Mycedium elephatotus	6, 9, 15, 25, 32, 33, 36–42, 46, 47
Mycedium mancaoi	1, 9, 12, 19, 24, 25, 31, 32, 34, 40–42, 45–47
Mycedium robokaki	1, 7, 9, 11, 12, 14–17, 20, 21, 23–25, 29, 33–37, 40–43, 46, 47
Oxypora crassispinosa	1, 7, 15, 19, 20, 24, 29, 31, 33, 34, 36, 38, 40–47
Oxypora glabra	2, 4, 5, 22, 23, 35

Species	Site Records
Oxypora lacera	1, 3, 5, 9, 13, 18–21, 23–25, 29, 30, 33, 34, 35, 37–43, 45–47
Pectinia alcicornis	23
Pectinia elongata	4, 18, 22, 44
Pectinia lactuca	1, 3, 4, 6–10, 12, 14–16, 20, 21, 23–25, 29–33, 35–42, 44–47
Pectinia maxima	44
Pectinia paeonia	2, 3, 7, 11, 20, 23, 29, 31, 35, 41, 43, 44, 46, 47
Pectinia teres	2, 10, 18, 29, 31, 43, 44, 46

Family Mussidae

Australomussa rowleyensis	1, 3, 15, 19, 22, 23, 25, 38, 44, 47
Cynarina lacrymalis	2, 18, 20, 22, 37, 38
Scolymia vitiensis	3, 4, 7, 9, 10, 12, 15–17, 19–25
Acanthastrea echinata	1, 3, 6, 7, 10, 19, 20, 33, 41
Lobophyllia corymbosa	5, 8, 9, 11, 17, 20–23, 29, 40, 44, 45
Lobophyllia flabelliformis	5, 16, 21, 23, 31, 36–38, 40, 44
Lobophyllia hataii	2, 3, 9, 24, 32
Lobophyllia hemprichii	2–22, 24, 25, 29, 31, 32, 34, 36–44, 46, 47
Lobophyllia robusta	2, 5, 7, 9, 13, 16, 17, 20–23, 29–34, 36–42, 44, 46
Symphyllia agaricia	1, 7, 8, 14–16, 19–21, 23, 25, 29–33, 37, 38, 40–43, 46, 47
Symphyllia hassi	4, 5, 17, 19–22, 24, 25, 42, 46, 47
Symphyllia radians	1, 12, 15, 25, 30, 33, 36, 37, 39, 43, 46, 47
Symphyllia recta	5, 7–9, 12, 14–16, 18, 20–23, 29–32, 36–42, 44–47
Symphyllia valenciennesii	8, 15, 20, 22

Family Merulinidae

Hydnophora exesa	3–5, 7, 9, 12, 14–17, 21, 29, 30, 32–34, 37, 38, 40–42, 45
Hydnophora grandis	6, 9, 10, 12–16, 21, 22, 24, 29, 32, 36, 38, 40, 44–47
Hydnophora microconos	13, 25, 30, 31, 33, 34, 40–42, 46, 47
Hydnophora pilosa	7, 34, 36, 38, 47
Hydnophora rigida	1–4, 7, 14, 18, 23, 29–34, 36, 40–43, 45–47
Merulina ampliata	1–5, 7, 9–25, 29–31, 35, 37–42, 44–47
Merulina scabricula	1, 2, 6, 9, 10, 12, 14–17, 21, 22, 24, 25, 29–32, 34, 37–40, 43–47
Scapophyllia cylindrica	6, 7, 12, 15, 22, 24, 37, 41, 44, 47

Family Faviidae

Caulastrea echinulata	22, 34, 39, 46
Caulastrea furcata	34, 36
Favia laxa	12, 14, 16, 24, 45–47
Favia maxima	11, 18, 36
Favia pallida	2, 3, 5–10, 13, 16, 19–22, 24, 25, 29–34, 36, 38–44, 46, 47
Favia rotundata	5, 8, 9, 38, 40, 43, 44
Favia speciosa	10, 16, 20, 31, 32, 34, 36–41, 43, 46
Favia stelligera	9, 10, 12–16, 19, 21, 24, 29, 39–42, 46, 47
Favia truncatus	36, 40, 41
Barabattoia amicorum	22
Favites abdita	5, 9, 10, 12–14, 16, 19, 20, 24, 32, 34, 36, 38–41, 43, 45–47
Favites flexuosa?	9, 15, 18, 22, 24, 30, 32–34, 36, 44–46

Species	Site Records
Favites halicora	7, 17, 21, 36
Favites pentagona	16, 37, 46
Goniastrea australensis?	8, 9, 13, 18
Goniastrea edwardsi	4, 8, 11–13, 24, 32, 38, 40
Goniastrea pectinata	2–6, 9, 10, 12–14, 16, 17, 19–22, 24, 25, 30–33, 36, 37, 42–46
Goniastrea retiformis	2, 5, 6, 9, 12–14, 16, 17, 19, 20, 24, 25, 30, 38–47
Platygyra acuta	5
Platygyra daedalea	1, 6, 7, 9, 12, 13, 15, 16, 20, 22, 24, 25, 29, 30, 32–34, 36–38, 40–47
Platygyra lamellina	1–3, 5, 7, 8, 23, 25, 30, 35, 37–39, 43
Platygyra pini	3, 5–10, 13–18, 20–23, 25, 29–31, 33, 36–38, 40–44, 46
Platygyra sinensis	6, 9, 29, 30, 36, 38–40, 42, 43, 45
Platygyra verweyi	21
Plesiastrea versipora	16, 20, 24, 30, 38, 41, 46
Leptoria irregularis	
Leptoria phrygia	1, 5, 7, 9, 13, 15, 24, 30, 33, 34, 36, 37, 40, 45–47
Oulophyllia bennettae	7, 11, 20, 33, 41
Oulophyllia crispa	7, 8, 25, 30, 32–34, 40, 43, 45, 47
Montastrea annuligera	7
Montastrea colemani	19, 44
Montastrea curta	3, 13, 42
Montastrea magnistellata	3, 44
Montastrea salebrosa	3, 10, 11, 16, 17, 24, 43
Oulastrea crispata	13, 40
Diploastrea heliopora	1–4, 6, 8–11, 13–25, 29–32, 34, 36–46
Leptastrea bewickensis	22
Leptastrea pruinosa	2, 11, 15, 20, 22, 24, 33, 36
Leptastrea purpurea	1, 3–6, 9–11, 14–22, 24, 32, 38, 40, 46, 47
Leptastrea transversa	7, 15, 16, 25, 32, 38, 39
Cyphastrea decadia	12, 22, 29, 42
Echinopora ashmorensis	1, 6, 8, 12, 21, 24, 42, 44, 46, 47
Echinopora gemmacea	3, 5, 12, 15–17, 19, 21, 24, 25, 31, 32, 41, 44
Echinopora hirsuitissima	7, 12–15, 24, 25, 45
Echinopora horrida	6
Echinopora lamellosa	1, 6, 7, 14, 15, 17, 20, 21, 24, 29–38, 40–47
Echinopora mammiformis	3–5, 11, 17–19, 21–24, 40, 45, 46
Echinopora pacificus	3, 4, 6, 10, 21, 22, 25, 42, 46

Family Trachyphyllidae

Trachyphyllia geoffroyi	11, 18

Family Caryophilliidae

Euphyllia ancora	30, 31, 38–40, 43
Euphyllia divisa	30, 32, 36
Euphyllia glabrescens	4, 9, 18, 20, 22, 23, 25, 29–33, 36–38, 40, 43, 45, 46
Euphyllia paraancora	30, 32–34, 38
Euphyllia paradivisa	31, 34, 35
Euphyllia yaeyamaenisis	2, 23, 24, 30, 31, 37, 40, 42, 44
Plerogyra simplex	2–5, 7–10, 12, 14–24, 29, 31, 35, 37, 39–44, 46

Species	Site Records
Plerogyra sinuosa	23, 29, 31–33, 35–42, 47
Physogyra lichentensteini	1–3, 6, 8, 9, 11–18, 20–22, 24, 25, 29, 30, 33, 36–40, 42, 43, 46
Thalamophyllia tenuescens	47
Family Dendrophylliidae	
Turbinaria frondens	32, 37, 40, 44
Turbinaria irregularis	8, 15, 23, 46
Turbinaria mesenterina	8, 11, 15, 19–22, 25, 31, 32, 36, 38, 41–44, 46
Turbinaria peltata	1, 7, 16, 20, 33, 35, 37, 39, 40, 46, 47
Turbinaria reniformis	7, 9, 12, 14, 15, 19, 20, 22, 23, 25, 30, 31, 33, 35–38, 40–45, 47
Turbinaria stellulata	47
Dendrophyllia sp. 1	47
Dendrophyllia coccinea	38, 47
Tubastraea coccinea	30, 36, 47
Tubastraea faulkneri	36, 47
Tubastraea micranthus	7, 9, 15, 16, 30, 32, 33, 37, 40, 41, 47
Rhizopsammia verrilli	9, 12–15, 25, 30, 34, 36, 40, 47
Family Heliporidae	
Heliopora coerulea	5–7, 9, 10, 13, 20, 21, 24, 25, 30, 36–40, 44, 46
Heliopora sp. 1	7, 15, 32, 39, 40, 43, 44, 46
Family Clavulariidae	
Tubipora musica	9, 23, 25, 29, 33, 36–46
Tubipora sp. 1	30
Tubipora sp. 2	33, 39, 40
Family Milleporidae	
Millepora dichotoma	1, 3, 6, 9, 12, 14, 16, 18–20, 22, 23, 31, 39, 41–43, 45
Millepora exaesa	1, 3–6, 9, 12, 14–17, 19–22, 30, 33, 42, 44–47
Millepora intricata	6, 12–15, 20, 29, 30, 32, 37, 39, 41–45
Millepora platyphylla	1, 6, 9, 12–16, 20–22, 24, 25, 29, 30, 32, 33, 36, 39–41, 43, 44, 46
Family Stylasteridae	
Stylaster sp. 1	7, 9, 25, 30, 33, 36, 38, 43
Stylaster sp. 2	33, 37, 45, 47
Distichopora nitida	33, 37–40

Appendix 2

Data used for calculating the
Reef Condition Index (RCI)

Site	Coral Species	Fish Species	Condition Points	RCI
1	64	178	110	162.80
2	52	134	110	137.86
3	69	154	150	176.26
4	52	133	120	142.05
5	64	155	100	150.04
6	57	179	90	148.68
7	68	181	160	189.68
8	54	145	110	143.32
9	87	169	180	209.10
10	50	208	160	185.47
11	44	70	110	108.85
12	69	161	190	196.94
13	67	230	170	210.95
14	70	175	140	179.98
15	84	202	170	186.76
16	62	208	170	199.25
17	51	162	170	174.36
18	51	149	130	151.54
19	73	163	130	173.46
20	83	216	140	204.63
21	84	174	120	181.30
22	80	106	110	149.40
23	57	136	110	142.42
24	93	184	160	209.98
25	76	266	170	230.73
26	75	170	180	200.22
27	75	193	140	190.26
28	75	176	180	202.37
29	61	188	210	209.52
30	69	181	190	204.08
31	70	139	140	167.13

Site	Coral Species	Fish Species	Condition Points	RCI
32	71	177	200	208.74
33	69	215	140	193.50
34	56	151	140	160.64
35	29	120	170	142.44
36	67	174	110	163.68
37	64	213	150	193.48
38	86	210	120	195.70
39	60	175	110	158.65
40	92	195	110	190.41
41	87	183	90	173.19
42	89	189	110	185.96
43	73	147	90	149.56
44	87	138	100	161.66
45	89	160	180	207.42
46	100	184	210	238.09
47	83	197	170	211.48

Appendix 3

Percentage of various bottom cover at individual sites

Station	Depth	Hard Corals	Dead Corals	Rubble	Sand	Soft Coral	Sponges	Algae	Other Organisms	Total
Station 1										
	4–6	62	10	17	0	0	0	11	0	100.0
	12–16	31	6	55	1	0	7	0	0	100.0
	21–23	19	4	56	2	1	15	3	0	100.0
	Average	37	7	43	1	0	7	4	—	100.0
Station 2										
	4–6	61	23	9	1	0	5	0	1	100.0
	12–14	54	21	17	2	0	6	0	0	100.0
	23–26	29	15	22	15	0	10	7	2	100.0
	Average	48	20	16	6	—	7	2	1	100.0
Station 3										
	4–6	54	23	10	7	0	5	0	1	100.0
	12–14	48	28	10	13	0	0	0	1	100.0
	21–22	29	7	8	43	1	10	2	0	100.0
	Average	43.7	19.3	9.3	21.0	0.3	5.0	1.0	1.0	100.0
Station 4										
	4–6	31	16	9	20	0	13	11	0	100.0
	12–15	34	37	23	0	0	3	3	0	100.0
	23–25	44	32	13	0	0	9	1	1	100.0
	Average	36.3	28.3	15.0	6.7	—	8.3	5.0	0.3	100.0
Station 5										
	4–5	16	22	16	31	0	8	0	7	100.0
	12–15	54	30	2	8	0	4	2	0	100.0
	23–25	46	42	5	0	0	2	5	0	100.0
	Average	38.7	31.3	7.7	13.0	—	4.7	2.3	2.3	100.0

Station	Depth	Hard Corals	Dead Corals	Rubble	Sand	Soft Coral	Sponges	Algae	Other Organisms	Total
Station 6										
	4	14	47	0	20	9	4	0	6	100.0
	9	9	60	1	20	5	5	0	0	100.0
	Average	**11.5**	**53.5**	**0.5**	**20.0**	**7.0**	**4.5**	**—**	**3.0**	**100.0**
Station 7										
	4–8	57	30	6	0	2	4	0	1	100.0
	12–14	60	29	2	0	0	5	4	0	100.0
	23–26	48	29	3	0	6	14	0	0	100.0
	Average	**55.0**	**29.3**	**3.7**	**—**	**2.7**	**7.7**	**1.3**	**0.3**	**100.0**
Station 8										
	4	37	42	11	2	7	1	0	0	100.0
	8	21	27	33	12	0	7	0	0	100.0
	Average	**29.0**	**34.5**	**22.0**	**7.0**	**3.5**	**4.0**	**—**	**—**	**100.0**
Station 9										
	4–5	42	21	19	9	2	5	0	2	100.0
	23–25	56	22	7	1	3	10	0	1	100.0
	Average	**49.0**	**21.5**	**13.0**	**5.0**	**2.5**	**7.5**	**—**	**1.5**	**100.0**
Station 10										
	5–7	52	26	12	2	5	1	0	2	100.0
	10–12	42	34	0	18	0	3	1	2	100.0
	19–20	32	17	9	33	0	7	0	2	100.0
	Average	**42.0**	**25.7**	**7.0**	**17.7**	**1.7**	**3.7**	**0.3**	**2.0**	**100.0**
Station 11										
	3	30	32	14	10	0	4	9	1	100.0
	7	35	33	18	11	2	0	0	1	100.0
	Average	**32.5**	**32.5**	**16.0**	**10.5**	**1.0**	**2.0**	**4.5**	**1.0**	**100.0**
Station 12										
	4–6	53	16	24	0	7	0	0	0	100.0
	11–12	49	12	12	0	4	23	0	0	100.0
	18–21	73	6	6	1	0	13	0	1	100.0
	Average	**58.3**	**11.3**	**14.0**	**0.3**	**3.7**	**12.0**	**—**	**0.3**	**100.0**
Station 13										
	4–5	27	20	36	0	13	0	0	4	100.0
	9–13	37	11	2	0	10	36	0	4	100.0
	19–20	32	20	7	6	5	29	0	1	100.0
	Average	**32.0**	**17.0**	**15.0**	**2.0**	**9.3**	**21.7**	**—**	**3.0**	**100.0**

Station	Depth	Hard Corals	Dead Corals	Rubble	Sand	Soft Coral	Sponges	Algae	Other Organisms	Total
Station 14										
	9–12	58	10	32	0	0	0	0	0	100.0
	21–24	53	16	7	0	0	22	0	2	100.0
	Average	**55.5**	**13.0**	**19.5**	—	—	**11.0**	—	**1.0**	**100.0**
Station 15										
	4–5	58	24	12	0	4	2	0	0	100.0
	9–12	57	30	0	0	4	6	3	0	100.0
	18–21	56	15	7	0	0	19	3	0	100.0
	Average	**57.0**	**23.0**	**6.3**	—	**2.7**	**9.0**	**2.0**	**0**	**100.0**
Station 16										
	4–5	62	23	11	0	4	0	0	0	100.0
	11–12	33	24	0	0	1	24	18	0	100.0
	18–20	17	37	0	4	0	25	17	0	100.0
	Average	**37.3**	**28.0**	**3.7**	**1.3**	**1.7**	**16.3**	**11.7**	**0**	**100.0**
Station 17										
	5–6	65	21	12	0	0	2	0	0	100.0
	9–10	57	31	11	0	0	1	0	0	100.0
	21–18	65	28	4	0	0	3	0	0	100.0
	Average	**62.3**	**26.7**	**9.0**	**0**	**0**	**2.0**	**0**	**0**	**100.0**
Station 18										
	4–5	38	29	2	25	2	4	0	0	100.0
	13–14	26	48	4	18	4	0	0	0	100.0
	>20	0	0	0	100	0	0	0	0	100.0
	Average	**21.3**	**25.7**	**2.0**	**47.7**	**2.0**	**1.3**	—	—	**100.0**
Station 19										
	5–7	55	31	8	2	0	3	0	1	100.0
	13–15	47	42	7	2	0	2	0	0	100.0
	20	50	41	4	3	0	2	0	0	100.0
	Average	**50.7**	**38.0**	**6.3**	**2.3**	**0**	**2.3**	**0**	**0.3**	**100.0**
Station 20										
	5–6	35	40	9	8	0	4	0	4	100.0
	10–12	37	44	6	9	0	4	0	0	100.0
	20–23	54	29	5	6	0	6	0	0	100.0
	Average	**42.0**	**37.7**	**6.7**	**7.7**	—	**4.7**	**0**	**1.3**	**100.0**
Station 21										
	2–4	48	38	8	1	0	5	0	0	100.0
	8–10	47	25	9	5	0	14	0	0	100.0
	18–20	39	29	10	3	0	10	9	0	100.0
	Average	**44.7**	**30.7**	**9.0**	**3.0**	—	**9.7**	**3.0**	—	**100.0**

Station	Depth	Hard Corals	Dead Corals	Rubble	Sand	Soft Coral	Sponges	Algae	Other Organisms	Total
Station 22										
	5–6	53	29	5	9	0	4	0	0	100.0
	8–10	39	39	12	5	0	5	0	0	100.0
	18–20	21	50	9	18	0	2	0	0	100.0
	Average	**37.7**	**39.3**	**8.7**	**10.7**	**—**	**3.7**	**0**	**0**	**100.0**
Station 23										
	5–6	33	40	4	14	0	9	0	0	100.0
	10–11	29	49	10	5	0	7	0	0	100.0
	20–21	25	73	2	0	0	0	0	0	100.0
	Average	**29.0**	**54.0**	**5.3**	**6.3**	**—**	**5.3**	**—**	**—**	**100.0**
Station 24										
	5–6	53	23	19	0	2	3	0	0	100.0
	10–11	59	21	11	5	0	4	0	0	100.0
	19–20	57	17	18	0	0	6	1	1	100.0
	Average	**56.3**	**20.3**	**16.0**	**1.7**	**0.7**	**4.3**	**0.3**	**0.3**	**100.0**
Station 25										
	5–7	68	13	6	0	7	3	0	3	100.0
	15–17	59	21	6	5	4	5	0	0	100.0
	20–21	62	17	8	0	6	7	0	0	100.0
	Average	**63.0**	**17.0**	**6.7**	**1.7**	**5.7**	**5.0**	**—**	**1.0**	**100.0**
Station 26										
	4–5	27	20	0	0	53	0	0	0	100.0
	10–11	12	17	0	0	66	5	0	0	100.0
	18–20	23	12	7	10	42	0	0	6	100.0
	Average	**20.7**	**16.3**	**2.3**	**3.3**	**53.7**	**1.7**	**—**	**2.0**	**100.0**
Station 27										
	4–6	25	2	0	0	48	5	0	20	100.0
	10–11	36	4	0	2	34	13	0	11	100.0
	20–21	13	14	10	16	32	0	0	15	100.0
	Average	**24.7**	**6.7**	**3.3**	**6.0**	**38.0**	**6.0**	**—**	**15.3**	**100.0**
Station 28										
	10–11	16	5	0	33	40	0	0	6	100.0
	4–5	58	10	23	1	0	1	0	7	100.0
		—	—	—	—	—	—	—	—	—
Station 29										
	10	91	7	2	0	0	0	0	0	100.0
	20	96	3	1	0	0	0	0	0	100.0
	Average	**81.7**	**6.7**	**8.7**	**0.3**	**—**	**0.3**	**—**	**2.3**	**100.0**

Station	Depth	Hard Corals	Dead Corals	Rubble	Sand	Soft Coral	Sponges	Algae	Other Organisms	Total
Station 30										
	3–4	53	0	10	29	8	0	0	0	100.0
	10–11	100	0	0	0	0	0	0	0	100.0
	16–17	85	0	0	8	7	0	0	0	100.0
	Average	**79.3**	—	**3.3**	**12.3**	**5.0**	—	—	—	**100.0**
Station 31										
	0–4	0	0	0	100	0	0	0	0	100.0
	5–20	0	0	0	100	0	0	0	0	100.0
	>20	0	0	0	100	0	0	0	0	100.0
	Average	—	—	—	**100.0**	—	—	—	—	**100.0**
Station 32										
	2–4	87	1	10	2	0	0	0	0	100.0
	10–11	66	3	15	10	4	2	0	0	100.0
	20–21	10	0	25	50	4	8	0	3	100.0
	Average	**54.3**	**1.3**	**16.7**	**20.7**	**2.7**	**3.3**	—	**1.0**	**100.0**
Station 33										
	4–5	56	12	4	0	16	0	11	1	100.0
	10–11	34	10	43	0	6	1	0	6	100.0
	18–19	45	14	20	7	6	6	1	1	100.0
	Average	**45.0**	**12.0**	**22.3**	**2.3**	**9.3**	**2.3**	**4.0**	**2.7**	**100.0**
Station 34										
	5–6	64	2	12	2	15	0	0	5	100.0
	10–11	25	0	61	3	8	0	0	3	100.0
	19–20	24	0	0	14	62	0	0	0	100.0
	Average	**37.7**	**0.7**	**24.3**	**6.3**	**28.3**	—	—	**2.7**	**100.0**
Station 35										
	2–4	77	11	12	0	0	0	0	0	100.0
	10–11	87	13	0	0	0	0	0	0	100.0
	20–21	89	8	0	3	0	0	0	0	100.0
	Average	**84.3**	**10.7**	**4.0**	**1.0**	—	—	—	—	**100.0**
Station 36										
	2–4	58	7	0	2	4	0	29	0	100.0
	10–11	19	6	57	4	9	2	0	3	100.0
	12–20	0	0	100	0	0	0	0	0	100.0
	Average	**25.7**	**4.3**	**52.3**	**2.0**	**4.3**	**0.7**	**9.7**	**1.0**	**100.0**
Station 37										
	4–6	29	15	5	14	23	2	0	12	100.0
	10–11	41	5	4	4	28	11	1	6	100.0
	18–20	33	20	5	10	9	17	0	6	100.0
	Average	**34.3**	**13.3**	**4.7**	**9.3**	**20.0**	**10.0**	**0.3**	**8.0**	**100.0**

Station	Depth	Hard Corals	Dead Corals	Rubble	Sand	Soft Coral	Sponges	Algae	Other Organisms	Total
Station 38										
	2–4	27	8	8	6	23	0	6	22	100.0
	10–11	21	4	5	14	30	1	6	19	100.0
	19–20	19	9	4	32	25	6	0	5	100.0
	Average	**22.3**	**7.0**	**5.7**	**17.3**	**26.0**	**2.3**	**4.0**	**15.3**	**100.0**
Station 39										
	4–5	5	3	32	0	60	0	0	0	100.0
	10–11	7	12	27	23	30	0	0	1	100.0
	19–20	9	19	33	17	14	5	0	3	100.0
	Average	**7.0**	**11.3**	**30.7**	**13.3**	**34.7**	**1.7**	**—**	**1.3**	**100.0**
Station 40										
	2–4	24	13	12	19	30	0	0	2	100.0
	10–11	22	9	23	16	5	1	0	24	100.0
	19–20	9	0	10	71	1	3	0	6	100.0
	Average	**18.3**	**7.3**	**15.0**	**35.3**	**12.0**	**1.3**	**—**	**10.7**	**100.0**
Station 41										
	3–4	14	8	40	20	6	1	0	11	100.0
	10–11	17	12	7	28	21	4	0	0	89.0
	Average	**16.4**	**9.1**	**20.7**	**27.8**	**13.0**	**2.1**	**—**	**7.2**	**96.3**
Station 42										
	4–6	36	4	42	3	15	0	0	0	100.0
	10–11	14	5	59	14	2	6	0	0	100.0
	19–20	18	7	43	10	0	17	5	0	100.0
	Average	**22.7**	**5.3**	**48.0**	**9.0**	**5.7**	**7.7**	**1.7**	**—**	**100.0**
Station 43										
	2–4	65	9	12	12	0	0	0	2	100.0
	8–10	35	3	19	34	0	2	7	0	100.0
	19–20	18	18	60	2	0	2	0	0	100.0
	Average	**39.3**	**10.0**	**30.3**	**16.0**	**—**	**1.3**	**2.3**	**—**	**99.3**
Station 44										
	2–4	30	38	20	3	0	0	0	9	100.0
	10–11	22	9	42	22	0	3	0	2	100.0
	17–18	25	17	6	48	2	2	0	0	100.0
	Average	**25.7**	**21.3**	**22.7**	**24.3**	**0.7**	**1.7**	**—**	**3.7**	**100.0**
Station 45										
	3–4	89	4	0	0	7	0	0	0	100.0
	10–11	83	0	0	0	4	0	13	0	100.0
	16–17	83	8	0	0	0	0	9	0	100.0
	Average	**85.0**	**4.0**	**—**	**—**	**3.7**	**—**	**7.3**	**—**	**100.0**

Station	Depth	Hard Corals	Dead Corals	Rubble	Sand	Soft Coral	Sponges	Algae	Other Organisms	Total
Station 46										
	2–4	82	16	0	0	1	0	0	1	100.0
	10–11	74	8	4	6	2	6	0	0	100.0
	17–18	14	1	0	75	0	9	0	1	100.0
	Average	**56.7**	**8.3**	**1.3**	**27.0**	**1.0**	**5.0**	**—**	**0.7**	**100.0**
Station 47										
	2–4	58	33	7	2	0	0	0	0	100.0
	12–13	68	6	0	0	0	17	0	9	100.0
	23–26	62	8	0	0	0	23	0	7	100.0
	Average	**62.7**	**15.7**	**2.3**	**0.7**	**—**	**13.3**	**—**	**5.3**	**100.0**

Appendix 4

Mollusc species recorded in the Gulf of Tomini, Sulawesi, Indonesia

Species	Site Records
Class Polyplacophora	
Family Cryptoplacidae	
Cryptoplax larvaeformis (Burrow, 1815)	24
Family Ischnochitonidae	
Ischnochiton sp.	13, 18
Family Chitonidae	
Acanthopleura gemmata (Blainville, 1825)	20, 26, 27, 29
Tonicia lamellosa (Quoy & Gaimard, 1835)	18, 26
Class Gastropoda	
Family Patellidae	
Patella flexuosa Quoy & Gaimard, 1834	3, 20, 29
Patelloida saccharina (Linnaeus, 1758)	3, 8, 29
Patelloida striata (Quoy & Gaimard, 1834)	8
Family Haliotidae	
Haliotis asinina Linnaeus, 1758	2, 12, 31
Haliotis glabra Gmelin, 1791	6, 9, 12
Haliotis ovina Gmelin, 1791	1, 7, 8, 10, 14, 15, 24–26
Haliotis planata Sowerby, 1833	15
Haliotis sp. (juvenile)	2

Species	Site Records
Family Fissurellidae	
Diodora mus (Reeve, 1850)	8
Diodora galeata (Helbling, 1779)	13, 22, 29
Hemitoma panhi (Quoy & Gaimard, 1834)	12, 25
Family Turbinidae	
Astraea haemotragum (Menke, 1829)	20, 24, 26
Astralium calcar (Linnaeus, 1758)	2
Astralium rhodostomum (Lamarck, 1822)	15
Bolma erectospinosa (Habe & Okutani, 1980)	8, 20
Bolma persica (Dall, 1907)	30
Phasianella aff. *solida* (Born, 1778)	13
Turbo argyrostomus (Linnaeus, 1758)	6, 9, 13, 14, 17, 18, 24, 25, 31
Turbo bruneus (Röding, 1798)	22, 23
Turbo chrysostoma (Linnaeus, 1758)	5, 15, 16, 24–26
Turbo cinereus Born, 1778	8
Turbo aff. *foliaceus* Philippi, 1847	2, 3, 5, 6
Turbo petholatus Linnaeus, 1758	3–5, 17, 19, 21–23, 28, 30, 31
Liotina peronii (Kiener, 1839)	1
Family Trochidae	
Angaria delphinus (Linnaeus, 1758)	3, 6, 25
Astele pulcherrimus (Sowerby, 1914)	16, 21
Cantharidus gilberti (Montrouzier, 1878)	15, 23
Cantharidus picturatus (Adams, 1851)	7
Chrysostom paradoum (Born, 1780)	6, 8–10, 12
Clanculus clanguloides (Wood, 1828)	10, 14
Euchelus atratus (Gmelin, 1791)	2, 4
Herpetopoma atrata (Gmelin, 1791)	2, 20
Tectus conus (Gmelin, 1791)	29
Tectus fenestratus Gmelin, 1790	20
Tectus maculatus Linnaeus, 1758	2, 3, 6, 8, 9, 18, 20, 25, 29
Tectus niloticus Linnaeus, 1767	8, 9, 10, 16, 17, 25
Tectus pyramis Born, 1778	2, 3, 6, 17, 18, 20, 23, 24, 26
Tectus triserialis (Lamarck, 1822)	4, 21, 24, 31
Trochus stellatus Gmelin, 1791	29
Stomatella auricula (Lamarck, 1816)	12
Stomatella varia (Adams, 1850)	26
Stomatia phymotis Helbling, 1779	6, 22
Family Neritopsidae	
Neritopsis radula Gray, 1842	1, 5, 25
Family Neritidae	
Nerita albicilla Linnaeus, 1758	8, 20, 25
Nerita plicata Linnaeus, 1758	3, 8, 29
Nerita polita Linnaeus, 1758	20

Species	Site Records
Nerita reticulata Karsten, 1789	20, 25
Nerita undata Linnaeus, 1758	3, 8, 20, 29
Modulus tectum (Gmelin, 1791)	11, 20, 25, 26, 29

Family Cerithiidae

Species	Site Records
Cerithium balteatum Philippi, 1848	9, 12, 15
Cerithium batillariaeformis Habe & Kosuge, 1966	11
Cerithium columna Sowerby, 1834	6, 9, 12, 14, 15, 18, 23, 24
Cerithium echinatum (Lamarck, 1822)	9, 10, 16, 17, 19, 21, 25
Cerithium egenum Gould, 1849	12
Cerithium lifuense Melvill & Standen, 1895	17, 18, 20, 21
Cerithium moniliferum (Kiener, 1841)	1, 9, 20, 27
Cerithium nesioticum Pilsbry & Vanetta, 1906	8, 15, 16, 18
Cerithium nodulosus (Bruguière, 1792)	8
Cerithium planum Anton, 1839	6, 8, 18
Cerithium rostratum Sowerby, 1855	4, 8
Cerithium salebrosum Sowerby, 1855	2, 6, 8
Cerithium tenuifilosum Sowerby, 1866	16
Clypeomorus bifasciata (Sowerby, 1855)	20
Pseudovertagus aluco (Linnaeus, 1758)	18, 20, 22
Rhinoclavis articulata (Adams & Reeve, 1850)	8
Rhinoclavis aspera (Linnaeus, 1758)	2, 3, 5–10, 12, 16, 18, 20–26, 29–31
Rhinovlavis fasciatus (Bruguière, 1792)	3–6, 10, 15, 18, 26, 28
Rhinoclavis sinensis (Gmelin, 1791)	25, 31
Rhinoclavis vertagus (Linnaeus, 1767)	6, 25

Family Siliquariidae

Species	Site Records
Siliquaria sp.	5

Family Potamididae

Species	Site Records
Terebralia sulcata (Born, 1778)	20

Family Littorinidae

Species	Site Records
Littorina coccinea (Gmelin, 1791)	25, 29
Littorina scabra (Linnaeus, 1758)	6, 20, 25
Nodilittorina millegrana (Philippi, 1848)	3, 8
Tectarius coronatus Valenciennes, 1832	20
Tectarius grandinatus (Gmelin, 1791)	8

Family Strombidae

Species	Site Records
Lambis lambis (Linnaeus, 1758)	8
Lambis millepedes (Linnaeus, 1758)	1, 2, 4, 6, 8, 9, 12, 14, 16, 18–25, 27, 29, 30
Lambis scorpius (Linnaeus, 1758)	12, 24, 27
Strombus aurisdianae Linnaeus, 1758	20
Strombus dentatus Linnaeus, 1758	10, 12, 13, 21
Strombus gibberulus Linnaeus, 1758	4, 6, 8, 10, 12, 13, 18, 20

Species	Site Records
Strombus labiatus (Röding, 1798)	13
Strombus lentiginosus Linnaeus, 1758	4, 20, 25
Strombus luhuanus Linnaeus, 1758	8, 9, 16, 20, 25–27, 29, 31
Strombus microurceus (Kira, 1959)	13, 14
Strombus mutabilis Swainson, 1821	13, 20, 25
Strombus terebellatus Sowerby, 1842	3
Strombus urseus Linnaeus, 1758	2, 4, 11, 20, 23, 28, 29, 31
Strombus variabilis Swainson, 1820	1
Terebellum terebellum (Linnaeus, 1758)	2–6, 8, 10–12, 18, 22, 23, 26, 30

Family Hipponicidae

Hipponix conicus (Schumacher, 1817)	12, 13, 15

Family Capulidae

Cheilea equestris (Linnaeus, 1758)	3, 5, 14, 22, 23, 25
Crepidula walshi Reeve, 1859	13

Family Vermetidae

Serpulorbis colubrina (Röding, 1798)	1–3, 7, 9, 10, 14–21, 24–28

Family Cypraeidae

Cypraea annulus Linnaeus, 1758	8, 18, 25
Cypraea argus Linnaeus, 1758	6, 12
Cypraea asellus Linnaeus, 1758	6, 9, 11, 12, 15, 16, 20, 21, 23, 25, 27, 28
Cypraea caputserpentis Linnaeus, 1758	25, 27
Cypraea carneola Linnaeus, 1758	2, 3, 6, 7, 9, 11, 15, 23, 25–28, 30
Cypraea caurica Linnaeus, 1758	30
Cypraea chinensis Gmelin, 1791	27
Cypraea contaminata Sowerby, 1832	26
Cypraea cribraria Linnaeus, 1758	16, 26–28
Cypraea cylindrica Born, 1778	2, 4, 5, 8, 11, 18–21, 23–25, 27, 30
Cypraea eglantina (Duclos, 1833)	29
Cypraea erosa Linnaeus, 1758	2, 4, 22, 23, 25, 26, 29, 30
Cypraea fimbriata Gmelin, 1791	18, 26
Cypraea globulus Linnaeus, 1758	14
Cypraea helvola Linnaeus, 1758	25
Cypraea isabella Linnaeus, 1758	7, 20, 25, 26
Cypraea kieneri Hidalgo, 1906	5, 10
Cypraea labrolineata Gaskoin, 1848	3, 6, 15, 27
Cypraea lynx Linnaeus, 1758	1–4, 6, 10, 14–16, 21–23, 25, 29
Cypraea margarita Dillwyn, 1817	25
Cypraea microdon Gray, 1828	6, 9, 15, 22
Cypraea moneta Linnaeus, 1758	2, 6, 13–15, 18–22, 24–26, 30
Cypraea nucleus Linnaeus, 1758	9, 15, 25, 27
Cypraea pallidula Gaskoin, 1849	15
Cypraea punctata Linnaeus, 1758	20, 22
Cypraea quadrimaculata Gray, 1824	2

Species	Site Records
Cypraea staphylaea Linnaeus, 1758	8, 20, 25, 26, 30
Cypraea talpa Linnaeus, 1758	24, 25
Cypraea teres Gmelin, 1791	6, 16
Cypraea testudinaria Linnaeus, 1758	12, 14
Cypraea tigris Linnaeus, 1758	1, 6, 9, 11, 12, 15, 16, 18, 21, 22, 24, 27, 30
Cypraea ursellus Gmelin, 1791	30
Cypraea vitellus Linnaeus, 1758	3, 4, 31

Family Ovulidae

Species	Site Records
Calpurnus lacteus (Lamarck, 1810)	26
Ovula ovum (Linnaeus, 1758)	26
Testudovula nebula (Azuma & Cate, 1971)	26

Family Triviidae

Species	Site Records
Trivia oryza (Lamarck, 1810)	2, 3, 6, 9, 13, 15, 18, 19, 21–23, 26–28

Family Velutinidae

Species	Site Records
Coriocella nigra Blainville, 1824	20

Family Naticidae

Species	Site Records
Natica euzona (Récluz, 1844)	8
Natica gualtieriana (Récluz, 1844)	12
Natica onca (Röding, 1798)	6, 7, 18
Polinices aurantius (Röding, 1798)	31
Polinices sebae (Récluz, 1844)	18, 20
Polinices sordidus (Swainson, 1821)	12
Polinices tumidus (Swainson, 1840)	6, 20, 25, 29

Family Bursidae

Species	Site Records
Bursa granularis (Röding, 1798)	17, 25
Bursa lamarckii (Deshayes, 1853)	12, 14, 23, 30
Bursa rosa Perry, 1811	26, 29
Bursa tuberossissima (Reeve, 1844)	25
Tutufa bubo (Linnaeus, 1758)	15
Tutufa rubeta (Linnaeus, 1758)	10

Family Cassidae

Species	Site Records
Casmaria erinaceus (Linnaeus, 1758)	14, 18, 25, 30

Family Ranellidae

Species	Site Records
Charonia tritonis (Linnaeus, 1758)	10
Cymatium aquatile (Reeve, 1844)	26
Cymatium exile (Reeve, 1844)	12
Cymatium hepaticum Röding, 1798	9
Cymatium mundum (Gould, 1849)	26
Cymatium nicobaricum (Röding, 1798)	15
Cymatium pileare (Linnaeus, 1758)	5, 14, 20
Cymatium rubeculum (Linnaeus, 1758)	14

Species	Site Records
Distorsio anus (Linnaeus, 1758)	8, 20, 21, 25
Gyrineum bituberculare (Lamarck, 1816)	6
Gyrineum gyrineum (Linnaeus, 1758)	26, 27, 29
Gyrineum pusillum (A. Adams, 1854)	26
Linatella succincta (Linnaeus, 1771)	3
Septa gemmata (Reeve, 1844)	15, 21, 23, 25
Family Tonnidae	
Malea pomum (Linnaeus, 1758)	3, 8, 12, 16, 25
Tonna cepa (Röding, 1798)	13
Tonna galea (Linnaeus, 1758)	
Tonna perdix (Linnaeus, 1758)	6, 8, 18
Family Epitoniidae	
Epitonium scalare (Linnaeus, 1758)	13
Family Eulimidae	
Thyca crystallina (Gould, 1846)	6, 13, 26
Mucronalia gigas Kuroda & Habe, 1950	8
Family Muricidae	
Chicoreus banksii (Sowerby, 1841)	26
Chicoreus brunneus (Link, 1810)	4, 6, 8, 16, 18, 22, 27–29, 31
Chicoreus microphyllus (Lamarck, 1816)	25
Chicoreus penchinati (Crosse, 1861)	18
Favartia rosamiae (D'Attilio & Myers, 1985)	18
Murex ramosus (Linnaeus, 1758)	8
Murex tenuirostrum Lamarck, 1822	26
Naquetia capucina Lamarck, 1822	20
Pterynotus barclayanus (A. Adams, 1873)	7, 19, 25
Cronia funiculus (Wood, 1828)	20, 25
Cronia margariticola (Broderip, 1833)	8, 25
Drupa grossularia (Röding, 1798)	6, 8, 14, 16, 22, 25, 29, 30
Drupa morum (Röding, 1798)	27
Drupa ricinus (Linnaeus, 1758)	15, 24–27, 29
Drupa rubusidaeus (Röding, 1798)	14, 16, 21, 23–26
Drupella cariosa (Wood, 1828)	2–6, 14, 16, 20, 22, 24, 29, 31
Drupella cornus (Röding, 1798)	1, 2, 6, 7, 9, 12, 16, 17, 19, 21, 22, 24, 27, 29, 30
Drupella ochrostoma (Blainville, 1832)	9, 12, 16, 19–21, 25–31
Drupella rugosa (Born, 1778)	11
Maculotriton serriale (Deshayes, 1831)	13
Morula anaxeres (Kiener, 1835)	18, 25–27
Morula aurantiaca (Hombron & Jacquinot, 1853)	20
Morula granulata (Duclos, 1832)	27, 29
Morula fiscella (Gmelin, 1791)	5
Morula spinosa (H. & A. Adams, 1855)	6, 15, 18, 21, 22, 28, 29
Morula uva (Röding, 1798)	1, 6, 8, 9, 12, 15, 18, 19, 22, 24, 25, 27, 29
Thais kieneri (Deshayes, 1844)	19

Species	Site Records
Thais mancinella (Linnaeus, 1758)	1, 2, 26, 27
Thais tuberosa (Röding, 1798)	27
Coralliophila erosa (Röding, 1798)	9
Coralliophila violacea (Kiener, 1836)	2, 3, 5, 7–12, 14–17, 19–25, 27, 29–31
Quoyola madreporarum (Sowerby, 1832)	9, 25
Rapa rapa (Gmelin, 1791)	6, 12, 26, 27, 29, 30

Family Vasidae

Vasum ceramicum (Linnaeus, 1758)	9, 28
Vasum turbinellus (Linnaeus, 1758)	5, 6, 10, 14, 16, 29, 31

Family Buccinidae

Colubraria castanea Kuroda & Habe, 1952	30
Colubraria muricata (Lightfoot, 1786)	6, 7
Colubraria nitidula (Sowerby, 1833)	21, 25, 27, 28
Colubraria tortuosa (Reeve, 1844)	24
Colubraria sp.	24, 25
Cantharus fumosus (Dillwyn, 1817)	2, 6, 11, 20
Cantharus pulcher (Reeve, 1846)	12, 17, 25–29
Cantharus subrubiginosus (E.A. Smith, 1879)	18, 19, 21
Cantharus undosus (Linnaeus, 1758)	7, 16, 24–29
Cantharus wagneri (Anton, 1839)	26
Engina alveolata (Kiener, 1836)	9, 18, 20, 29
Engina egregialis (Reeve, 1844)	26
Engina incarnata (Deshayes, 1834)	1, 6, 7, 15
Engina lineata (Reeve, 1846)	20, 24, 29, 30
Engina zonalis (Lamarck, 1822)	20
Phos textum (Gmelin, 1791)	3, 4, 6, 12, 18, 20, 22, 29–31
Pisania fasciculata (Reeve, 1846)	25
Pisania gracilis (Reeve, 1846)	12
Pisania ignea (Gmelin, 1791)	14, 26

Family Columbellidae

Mitrella ligula (Duclos, 1840)	8, 11, 12, 15, 18, 20, 25–27
Pyrene deshayesii (Crosse, 1859)	4, 5, 7, 8, 11, 12, 15–20, 22, 23
Pyrene flava (Bruguière, 1789)	14
Pyrene punctata (Bruguière, 1789)	7, 15, 27, 29, 30
Pyrene scripta (Lamarck, 1822)	23–25, 28
Pyrene testudinaria (Link, 1807)	20
Pyrene turturina (Lamarck, 1822)	1, 2, 16, 17

Family Nassariidae

Nassarius albescens (Dunker, 1846)	2, 13, 18
Nassarius burchardi (Dunker, 1849)	13
Nassarius comptus (Adams, 1852)	22
Nassarius crematus (Hinds, 1844)	14
Nassarius ecstilbus (Melvill & Standen, 1896)	18
Nassarius glans (Linnaeus, 1758)	1

Species	Site Records
Nassarius granifeus (Kiener, 1834)	10, 15, 16, 25
Nassarius multicostatus A. Adams, 1852	13
Nassarius pauperus (Gould, 1850)	26
Nassarius pullus (Linnaeus, 1758)	13
Nassarius sp.	13
Family Fasciolariidae	
Dolicholatirus lancea (Gmelin, 1791)	18, 21, 25
Latirolagena smaragdula (Linnaeus, 1758)	6, 26, 27
Latirus belcheri (Reeve, 1847)	2, 10, 16, 19, 23
Latirus craticularis (Linnaeus, 1758)	24
Latirus gibbulus (Gmelin, 1791)	27, 28
Latirus nodatus (Gmelin, 1791)	1, 6, 9, 10, 14–16, 27
Latirus pictus (Reeve, 1847)	6, 7, 9, 10, 14–16, 18, 24
Latirus turritus (Gmelin, 1791)	1, 10, 12, 14–16, 19, 21, 24, 26–29
Peristernia hesterae Melvill, 1911	17
Peristernia incarnata (Deshayes, 1830)	17, 26–30
Peristernia nassatula (Lamarck, 1822)	20, 23, 25, 29
Peristernia ustulata (Reeve, 1847)	7
Pleuroploca filamentosa (Röding, 1798)	7, 10, 25, 29
Family Volutidae	
Cymbiola aulica (Sowerby, 1825)	5, 18, 22, 27, 28
Cymbiola rutila (Broderip, 1826)	11–13
Cymbiola vespertilio (Linnaeus, 1758)	9
Family Olividae	
Oliva annulata (Gmelin, 1791)	1, 2, 6, 7, 9, 10, 12, 16, 20–22, 25–27, 29, 30
Oliva carneola (Gmelin, 1791)	4–6, 8, 12–14, 21
Oliva miniacea Röding, 1798	29
Oliva oliva Linnaeus, 1758	26
Oliva parkinsoni Prior, 1975	28
Oliva tessellata Lamarck, 1811	10, 13, 18, 20, 21
Olivella sp.	2–4
Family Mitridae	
Cancilla filaris (Linnaeus, 1771)	5, 22
Imbricaria conovula (Quoy & Gaimard, 1833)	20
Imbricaria conularis (Lamarck, 1811)	10
Imbricaria olivaeformis (Swainson, 1821)	7, 12, 16, 24–25
Imbricaria punctata (Swainson, 1821)	8–10, 16
Imbricaria vanikorensis (Quoy & Gaimard, 1833)	18
Mitra avenacea Reeve, 1845	5
Mitra coarctata Reeve, 1844	12
Mitra contracta Swainson, 1820	3, 7, 8, 12, 14, 21, 22
Mitra cucumerina Lamarck, 1811	6, 10
Mitra fraga (Quoy & Gaimard, 1833)	26
Mitra luctuosa A. Adams, 1853	30

Species	Site Records
Mitra rubritincta Reeve, 1844	19, 23
Mitra ustulata Reeve, 1844	19
Neocancilla clathrus (Gmelin, 1791)	12, 25, 27, 28
Pterygia fenestrata (Lamarck, 1811)	11
Pterygia scabricula (Linnaeus, 1758)	30

Family Costellariidae

Vexillum aureolineatum Turner, 1988	12, 20
Vexillum cadaverosum (Reeve, 1844)	12, 13, 25
Vexillum consanguineum (Reeve, 1845)	10, 16, 18
Vexillum coronatum (Helbling, 1779)	12, 17
Vexillum costatum (Gmelin, 1791)	18
Vexillum deshayesii (Reeve, 1844)	13
Vexillum dennisoni (Reeve, 1844)	12
Vexillum discolorium (Reeve, 1845)	20
Vexillum echinatum (A. Adams, 1853)	13
Vexillum exasperatum (Gmelin, 1791)	4, 12, 13, 20
Vexillum granosum (Gmelin, 1791)	8, 18
Vexillum lucidum (Reeve, 1845)	20
Vexillum leucodesmium (Reeve, 1845)	15, 20
Vexillum militaris (Reeve, 1845)	26
Vexillum pacificum (Reeve, 1845)	12, 15
Vexillum plicarium (Linnaeus, 1758)	20
Vexillum polygonum (Gmelin, 1791)	13
Vexillum cf rubrum (Broderip, 1836)	18
Vexillum sanguisugum (Linnaeus, 1758)	5, 6, 10, 13
Vexillum semifasciatum (Lamarck, 1811)	13
Vexillum stainforthi (Reeve, 1841)	9, 13
Vexillum tankervillei (Melvill, 1888)	26
Vexillum turrigerum (Reeve, 1845)	20
Vexillum unifasciatus (Wood, 1828)	21
Zierliana anthricina (Reeve, 1844)	5

Family Turridae

Clavus canalicularis (Röding, 1798)	25, 31
Clavus flammulatus (Montfort, 1810)	8
Clavus unizonalis (Lamarck, 1822)	10, 26
Clavus viduus Reeve, 1845	18
Eucithara reticulata (Reeve, 1846)	11
Gemmula sp.	22
Lophiotoma acuta (Perry, 1811)	8
Lophiotoma albina (Lamarck, 1822)	11, 12
Splendrillia sp.	31
Turridrupa cerithina (Anton, 1839)	28
Turris babylonia (Linnaeus, 1758)	12
Xenoturris cingulifera (Lamarck, 1822)	10, 12, 28

Species	Site Records
Family Terebridae	
Duplicaria raphanula (Lamarck, 1822)	26
Hastula albida (Gray, 1834)	13
Hastula lanceata (Linnaeus, 1767)	12, 16, 28
Terebra affinis Gray, 1834	6, 7, 9, 10, 12, 13, 16, 20, 25, 28, 20
Terebra anilis Röding, 1798	13
Terebra argus Hinds, 1844	30
Terebra babylonia Lamarck, 1822	10, 22, 23
Terebra columellaris Hinds, 1844	16
Terebra crenulata (Linnaeus, 1758)	15, 25, 30
Terebra cumingi Deshayes, 1857	3, 8, 27, 28
Terebra felina (Dillwyn, 1817)	1-, 15, 25, 27, 28
Terebra funiculata Hinds, 1844	30
Terebra maculata (Linnaeus, 1758)	·7–10, 13, 15, 16, 25
Terebra punctostriata Gray, 1834	18
Terebra quoygaimardi Cernohorsky & Bratcher, 1976	11
Terebra succincta (Gmelin, 1791)	13
Family Conidae	
Conus arenatus Hwass in Bruguière, 1792	1–4, 10, 12–14, 16, 20–23, 25, 26, 29
Conus aurisiacus Linnaeus, 1758	14
Conus boeticus Reeve, 1844	8, 24, 26
Conus capitaneus Linnaeus, 1758	2, 23, 26, 27, 29
Conus catus Hwass in Bruguière, 1792	20
Conus circumactus Iredale, 1929	11
Conus circumcisus Born, 1778	9, 14, 15, 19
Conus coffeae Gmelin, 1791	9
Conus cf. comatosa Pilsbry, 1904	7
Conus coronatus (Gmelin, 1791)	26, 28
Conus distans Hwass in Bruguière, 1792	6, 26, 27
Conus emaciatus Reeve, 1849	8, 9, 14, 15
Conus flavidus Lamarck, 1810	3, 16, 23, 26, 27
Conus generalis Linnaeus, 1767	12
Conus geographus Linnaeus, 1758	9, 14, 19, 25
Conus glans Hwass in Bruguière, 1792	7, 25
Conus imperialis Linnaeus, 1758	9, 15, 23, 27
Conus litteratus Linnaeus, 1758	9, 12, 25, 28, 29
Conus lividus Hwass in Bruguière, 1792	22
Conus marmoreus Linnaeus, 1758	6, 8, 10, 15, 16, 18–21, 23, 25
Conus miles Linnaeus, 1758	1, 9, 13–16, 19, 21, 23, 25–28
Conus miliaris Hwass in Bruguière, 1792	15, 16, 26, 27
Conus monachus Linnaeus, 1758	8
Conus musicus Hwass in Bruguière, 1792	5–10, 12, 13, 15, 16, 18–20, 22–27, 30
Conus mustelinus Hwass in Bruguière, 1792	4, 6, 7, 11, 12, 14–19, 21, 23, 29–31
Conus nussatella Linnaeus, 1758	17, 24, 25
Conus parius Reeve, 1844	18
Conus pertusus Hwass in Bruguière, 1792*v*	26
Conus planorbis Born, 1778	3, 9, 18, 20, 21

Species	Site Records
Conus pulicarius Hwass in Bruguière, 1792	5
Conus radiatus Gmelin, 1791	10
Conus rattus Hwass in Bruguière, 1792	15, 17, 21
Conus sanguinolentus Quoy & Gaimard, 1834	26, 27
Conus spectrum Linnaeus, 1758	25
Conus sponsalis Hwass in Bruguière, 1792	12, 15, 16, 23, 25–28
Conus stercmuscarum Linnaeus, 1758	6, 15
Conus cf. striatellus Link, 1807	8
Conus striatus Linnaeus, 1758	29, 30
Conus tessellatus Born, 1778	30
Conus textile Linnaeus, 1758	15, 27
Conus varius Linnaeus, 1758	16
Conus vexillum Gmelin, 1791	9, 15, 16
Conus viola Cernohorsky, 1977	25
Conus virgo Linnaeus, 1758	15
Conus sp.	13
Family Architectonicidae	
Philippia radiata (Röding, 1798)	12
Family Pyramidellidae	
Pyramidella sp.	14
Family Acteonidae	
Acteon virgatus (Reeve, 1842)	12
Family Hydatinidae	
Hydatina physis (Linnaeus, 1758)	6
Family Haminoeidae	
Atys cylindricus (Helbling, 1779)	1, 3, 5, 6, 8, 12, 13, 18
Atys naucum (Linnaeus, 1758)	1, 3–5, 13, 18, 22, 31
Family Bullidae	
Bulla vernicosa Gould, 1859	13
Family Plakobranchidae	
Plakobranchus ocellatus van Hasselt, 1824	3–5, 11, 21, 22, 29
Family Elysiidae	
Elysia sp.	4
Elysia ratna Marcus, 1965	2, 3
Thurdilla sp.	13
Family Aplysiidae	
Dolabella sp.	23

Species	Site Records
Family Dorididae	
Ardedoris egretta Rudman, 1984	8
Discodoris boholensis Bergh, 1877	29
Family Chromodorididae	
Chromodoris bullocki Collingwood, 1857	20, 22
Chromodoris elisabethina Bergh, 1877	1
Chromodoris lochi Rudman, 1982	3
Chromodoris sp.	15
Family Phyllidiidae	
Phyllidia coelestis Bergh, 1905	1, 5, 6, 23
Phyllidia elegans Bergh, 1869	1, 10, 24
Phyllidia aff. *nobilis* Bergh, 1869	20
Phyllidia pustulosa (Cuvier, 1804)	1, 3, 4, 6, 9, 11, 14, 15, 18, 23–25, 27, 28
Phyllidia varicosa Lamarck, 1801	11, 12, 23
Phyllidia sp. 1	10
Phyllidia sp. 2	6
Family Glaucidae	
Pteraeolidia ianthina (Angas, 1864)	13
Family Ellobiidae	
Cassidula nucleus (Gmelin, 1791)	20, 23, 26
Ellobium sp.	20
Ellobium aurisjudae Linnaeus, 1758	7
Family Siphonariidae	
Siphonaria javanica (Lamarck, 1819)	8
Siphonaria sirius Pilsbry, 1894	3
Family Onchidiidae	
Onchidium sp.	29
Class Bivalvia	
Family Mytilidae	
Lithophaga sp. 1	3, 5, 8, 19, 23–26
Lithophaga sp. 2	16
Modiolus philippinarum Hanley, 1843	18–20, 23, 26, 27, 29
Modiolus sp.	3, 8
Septifer bilocularis (Linnaeus, 1758)	11, 18, 20, 26

Species	Site Records
Family Arcidae	
Anadara maculosa (Reeve, 1844)	3, 11, 20, 22, 25
Barbatia amygdalumtotsum (Röding, 1798)	1–11, 16, 18, 21, 25, 30, 31
Barbatia foliata Forsskål, 1775	3, 6, 13, 14, 16, 18, 24–27, 29
Barbatia ventricosa (Lamarck, 1819)	3–7, 9–11, 15, 17, 19–27, 31
Trisidos semitorta (Lamarck, 1819)	11
Trisidos tortuosa (Linnaeus, 1758)	11
Arcid sp.	18, 20
Family Glycymerididae	
Tucetona amboiensis (Gmelin, 1791)	2, 3, 5, 9, 23, 28, 30, 31
Family Pteriidae	
Pinctada margaritifera (Linnaeus, 1758)	3, 9, 13, 25, 26, 29
Pteria pengiun (Röding, 1798)	3, 4, 6, 13, 18, 23
Family Malleidae	
Malleus anatina (Gmelin, 1791)	3
Malleus malleus (Linnaeus, 1758)	27, 31
Vulsella vulsella (Linnaeus, 1758)	6
Family Isognomonidae	
Isognomon sp.	2
Family Pinnidae	
Atrina vexillum (Born, 1778)	5
Pinna bicolor (Gmelin, 1791)	22, 26
Streptopinna saccata (Linnaeus, 1758)	1, 2, 12, 16, 24–26, 28–30
Family Limidae	
Ctenoides annulata (Lamarck, 1819)	2, 3, 6, 8
Lima cf. *basilanica* (A. Adams & Reeve, 1850)	11
Lima fragilis (Gmelin, 1791)	7, 9, 14, 15, 18, 19, 26, 27
Lima lima (Link, 1807)	2–5, 7, 9–12, 15, 16, 18, 22, 23, 28–31
Lima orientalis (Adams & Reeve, 1850)	16, 25
Lima sp.	14
Family Ostreidae	
Alecryonella plicatula (Gmelin, 1791)	1, 4, 5, 7, 9–11, 16, 17, 19, 21, 22
Hyotissa hyotis (Linnaeus, 1758)	2, 19, 26, 27
Lopha cristagalli (Linnaeus, 1758)	7, 16, 19, 22, 23, 27
Lopha sp.	2, 7, 16, 27, 29–31
Saccostrea cf *cucullata* (Born, 1778)	3, 8, 27, 29
Saccostrea echinata (Quoy & Gaimard, 1835)	3
Saccostrea sp. 1	4, 5, 9
Saccostrea sp. 2	1, 20, 24–26, 28, 30

Species	Site Records
Family Pectinidae	
Chlamys corsucans (Hinds, 1845)	15
Chlamys lentiginosa (Reeve, 1865)	1, 4, 16, 18
Chlamys mollita (Reeve, 1853)	14, 17
Chlamys rastellum (Lamarck, 1819)	20
Chlamys squamata (Gmelin, 1791)	9
Chlamys squamosa (Gmelin, 1791)	2, 4, 16, 18, 19, 21, 22, 26, 28, 30, 31
Chlamys sp.	14
Comptopallium radula (Linnaeus, 1758)	2, 6, 11, 16, 18, 21, 22, 26, 29, 30
Gloripallium pallium (Linnaeus, 1758)	2, 4, 9, 10, 12
Laevichlamys limatula Reeve, 1853	14
Mirapecten rastellum (Lamarck, 1819)	9, 18
Pedum spondyloidaeum (Gmelin, 1791)	1–7, 9–12, 15, 17–19, 21–25
Semipallium luculentum (Reeve, 1853)	10, 22, 31
Semipallium tigris (Lamarck, 1819)	6, 7, 9, 14, 19, 21, 25, 26, 28–31
Family Spondylidae	
Spondylus candidus (Lamarck, 1819)	2, 5, 7, 9, 10
Spondylus multimuricatus Reeve, 1856	9
Spondylus sanguineus Dunker, 1852	1, 5, 7, 12, 16, 18, 19, 21, 22, 24, 26, 31
Spondylus sinensis Schreibers, 1793	6, 23, 25
Spondylus squamosus Schreibers, 1793	1–3, 11
Spondylus varians Sowerby, 1829	21, 23
Family Chamidae	
Chama brassica Reeve, 1846	7
Chama lazarus Linnaeus, 1758	4
Chama limbula (Lamarck, 1819)	6
Chama savigni Lamy, 1921	1, 22
Chama sp.	6, 24, 26, 27, 29, 31
Family Lucinidae	
Anodontia edentula (Linnaeus, 1758)	13
Anodontia pila (Reeve, 1850)	3–5
Family Fimbriidae	
Codakia tigerina (Linnaeus, 1758)	1, 5, 18, 25, 29, 30
Fimbria fimbriata (Linnaeus, 1758)	5, 6, 8, 18, 22, 25, 29
Family Galeommatidae	
Scintilla sp.	15
Family Carditidae	
Beguina semiorbiculata (Linnaeus, 1758)	3, 7, 16, 17–19, 21–23, 31
Cardita variegata Bruguière, 1792	3–5, 7, 8, 16–19
Megacardita aff *incrassata* (Sowerby, 1825)	2, 10, 21, 24–26, 28, 31

Species	Site Records
Family Cardiidae	
Acrosterigma cygnorum (Deshayes, 1855)	4
Acrosterigma aff. *dupuchense* (Reeve, 1845)	11, 18, 23
Acrosterigma elongata (Bruguière, 1789)	4
Acrosterigma luteomarginata (Voskuil & Onverwagt, 1991)	9
Acrosterigma reeveanum (Dunker, 1852)	11
Acrosterigma unicolor (Sowerby, 1834)	11, 21, 31
Acrosterigma sp.	8, 18, 21, 23, 25–28, 30, 31
Corculum cardissa (Linnaeus, 1758)	1
Fragum fragum (Linnaeus, 1758)	3, 5, 6, 8, 10, 18, 25
Fragum hemicardium (Linnaeus, 1758)	5
Fragum unedo (Linnaeus, 1758)	8, 18, 19, 21, 22, 29, 31
Fulvia aperta (Bruguière, 1789)	2, 6
Trachycardium alternatum (Sowerby, 1841)	2, 31
Trachycardium elongatum (Bruguière, 1789)	1, 11
Trachycardium enode (Sowerby, 1841)	3, 5, 9–15, 19, 20, 22, 23, 25–27, 29–31
Trachycardium orbita (Sowerby, 1833)	14, 15
Vepricardium multispinosum (Sowerby, 1841)	31
Family Tridacnidae	
Hippopus hippopus (Linnaeus, 1758)	2, 4, 5, 11, 12, 18, 20, 22–25, 29
Hippopus porcellanus Rosewater, 1982	2–4, 14, 17, 18, 22, 23
Tridacnea crocea Lamarck, 1819	2–6, 8–11, 14–16, 18, 19, 21–24, 27–30
Tridacna gigas (Linnaeus, 1758)	2, 17
Tridacna maxima (Röding, 1798)	1
Tridacna squamosa Lamarck, 1819	1–4, 6–26, 29–31
Family Mactridae	
Mactra sp.	11, 18, 20, 21
Family Solenidae	
Solen lamarckii Deshayes, 1839	11, 31
Family Tellinidae	
Strigilla tomlini Smith, 1915	12
Tellina exculta Gould, 1850	8, 18
Tellina gargadia Linnaeus, 1758	2, 6, 7, 12, 15, 16, 18
Tellina linguafelis Linnaeus, 1758	11, 12, 19
Tellina ovalis Sowerby, 1825	12
Tellina palatum (Iredale, 1929)	2–5
Tellina pretium Salisbury, 1934	12
Tellina rastellum Hanley, 1844	10, 16
Tellina rostrata Linnaeus, 1758	31
Tellina scobinata Linnaeus, 1758	2, 14, 18, 20, 22–25
Tellina staurella Lamarck, 1818	29

Species	Site Records
Family Semelidae	
Semele casta A. Adams, 1853	11, 22, 23
Semele duplicata (Sowerby, 1833)	26
Semele jukesii (Reeve, 1853)	15, 21
Semele lamellosa (Sowerby, 1830)	2
Family Psammobiidae	
Gari amethystus (Wood, 1815)	25, 26
Gari maculosa (Lamarck, 1818)	6
Gari occidens (Gmelin, 1791)	6
Gari pulcherrimus (Deshayes, 1855)	3
Gari squamosa (Lamarck, 1818)	18
Family Donacidae	
Donax sp.	12, 25
Family Trapeziidae	
Trapezium bicarinatum (Schumacher, 1817)	2
Trapezium obesa (Reeve, 1843)	22, 31
Family Veneridae	
Antigona clathrata (Deshayes, 1854)	16
Antigona chemnitzii (Hanley, 1844)	15
Antigona cf. *corbis* (Lamarck, 1818)	23
Antigona purpurea (Linnaeus, 1771)	2, 6
Antigona restriculata (Sowerby, 1853)	7, 9, 16, 21, 28
Antigona reticulata (Linnaeus, 1758)	3
Callista lilacina (Lamarck, 1818)	12
Callista sp.	12, 13
Dosinia amphidesmoides (Reeve, 1850)	26
*Dosinia incis*a (Reeve, 1850)	1
Dosinia juvenilis (Gmelin, 1791)	3, 4, 26
Dosinia sp.	12, 16, 25
Gafrarium tumidum Röding, 1798	2, 20
Globivenus toreuma (Gould, 1850)	1, 5, 7, 11, 14–17, 23, 25, 26, 28, 30
Lioconcha annettae Lamprell & Whitehead, 1990	3, 6, 15, 18, 30
Lioconcha castrensis (Linnaeus, 1758)	3, 4, 6, 7, 9, 10, 12, 16, 18, 20–23, 25, 31
Lioconcha fastigiata (Sowerby, 1851)	31
Lioconcha sp.	21
Lioconcha ornata (Dillwyn, 1817)	1, 11, 18, 31
Lioconcha polita (Röding, 1798)	8, 12
Paphia gallus (Gmelin, 1791)	11
Pitar affinis (Gmelin, 1791)	23
Pitar prora (Conrad, 1837)	3, 4
Placamen calophylla (Philippi, 1836)	11
Placamen tiara (Dillwyn, 1817)	10
Tapes sulcarius Lamarck, 1818	22, 26, 27, 29, 30
Timoclea marica (Linnaeus, 1758)	4, 18, 31

Species	Site Records
Family Corbulidae	
Corbula cf *taheitensis* Lamarck, 1818	12, 16, 30
Family Teredinidae	
Teredinid sp.	12, 13
Class Cephalopoda	
Family Sepiidae	
Sepia sp.	25, 29
Family Octopodidae	
Octopus sp.	10
Class Scaphopoda	
Family Dentaliidae	
Dentalium crocinum (Dall, 1907)	5, 11–13, 18
Dentalium elephantinum Linnaeus, 1758	3, 4

Appendix 5

Reef fishes recorded during the RAP survey of the Togean and Banggai Islands, Sulawesi, Indonesia

The phylogenetic sequence of the families appearing in this list follows Eschmeyer (Catalog of Fishes, California Academy of Sciences, 1998) with slight modification (e.g., placement of Cirrhitidae). Genera and species are arranged alphabetically within each family.

Terms relating to relative abundance are as follows:
Abundant—common at most sites in a variety of habitats with up to several hundred individuals being routinely observed on each dive.

Common—seen at the majority of sites in numbers that are relatively high in relation to other members of a particular family, especially if a large family is involved.
Moderately common—not necessarily seen on most dives, but may be relatively common when the correct habitat conditions are encountered.
Occasional—infrequently sighted and usually in small numbers, but may be relatively common in a very limited habitat.
Rare—less than 10, often only one or two individuals seen on all dives.

Species	Site Records	Abundance	Depth (m)
Ginglymostomatidae			
Nebrius ferrugineus (Lesson, 1830)	45	Rare, a single individual sighted.	1–70
Carcharhinidae			
Carcharhinus melanopterus (Quoy & Gaimard, 1824)	19	Rare, a single individual sighted.	0–10
Dasyatididae			
Dasyatis kuhlii (Müller & Henle, 1841)	27, 37, 40, 42	Rare, only four sighted.	2–50
Taeniura lymma (Forsskål, 1775)	25, 27–29, 31, 32, 35, 37–39, 42, 43, 46	Occasional.	2–30
Chlopsidae			
Kaupichthys brachychirus Shultz, 1953	43	Collected with rotenone at one site.	3–25
K. hypoproroides (Stromann, 1896)	13, 15	Several specimens collected with rotenone.	5–25
Muraenidae			
Echidna nebulosa (Thunberg, 1789)	34	A single individual observed.	1–10
Gymnothorax fimbriatus (Bennett, 1831)	40	A single individual observed.	0–30
G. flavimarginatus (Rüppell, 1828)	37, 38	Rare, only two seen.	1–150
G. javanicus (Bleeker, 1865)	1, 12, 13, 16, 24	Rare, only five seen.	0.5–50
G. melatremus Schultz, 1953	13, 15, 42	Several specimens collected with rotenone.	5–30
G. sp. 1	13	One specimen collected with rotenone.	10–20
G. sp. 2	37	One specimen collected with rotenone	10–25
G. zonipectus Seale, 1906	14, 43	Two specimens collected with rotenone.	8–45
Rhinomuraena quaesita Garman, 1888	20,46	Rare, only two seen.	1–50
Uropterygius concolor Rüppell, 1838	13	One specimen collected with rotenone.	0–20
U. kamar McCosker & Randall, 1977	13, 15	Several specimens collected with rotenone.	3–55
Ophichthidae			
Callechelyn marmoratus (Bleeker, 1852)	31	Rare, but difficult to detect.	1–15
Muraenichthys macropterus Bleeker, 1857	21	Two specimens collected with rotenone.	0–15

Species	Site Records	Abundance	Depth (m)
Congridae			
Gorgasia sp.	1, 7, 10, 13	Locally common, occurring in colonies with several hundred individuals.	15–35
Heteroconger haasi (Klausewitz and Eibesfeldt, 1959)	27, 28, 30, 32, 40	Locally common, occurring in colonies with several hundred individuals.	3–45
Clupeidae			
Spratelloides gracilis (Temminck & Schlegel, 1846)	2–4, 8	Schools occasionally encountered.	0–4
Plotosidae			
Plotosus lineatus (Thunberg, 1787)	38	Rare, one aggregation of about 50 small juveniles seen.	1–20
Synodontidae			
Saurida gracilis (Quoy and Gaimard, 1824)	45	One specimen collected with rotenone.	1–130
Synodus dermatogenys Fowler, 1912	12, 25, 27, 28, 31, 32, 34, 37, 40, 41, 43	Occasional.	1–25
S. jaculum Russell & Cressy, 1979	16, 26, 27, 39, 47	Rare.	2–85
S. variegatus (Lacepède, 1803)	2–4, 10, 20, 25, 31, 32, 35, 37–39, 41, 44, 45	Occasional.	5–50
Ophidiidae			
Brotula multibarbata (Temminck & Schlegel, 1846)	37	A single specimen collected with rotenone.	5–150
Carapidae			
Onuxodon margaritiferae (Rendahl, 1921)	13	Two specimens taken from a pearl–oyster shell.	5–30
Bythitidae			
Ogilbia sp. 1	13, 21, 34	Several specimens collected with rotenone.	10–25
O. sp. 2	15	Two specimens collected with rotenone.	10–25
Antennariidae			
Antennarius rosaceus Smith and Radcliffe, 1912	43	A single small juvenile collected with rotenone.	0–130
Atherinidae			
Atherinomorus lacunosus (Forster, 1801)	3, 4, 8, 16, 35	Occasional schools sighted.	0–3

Species	Site Records	Abundance	Depth (m)
Belonidae			
Tylosurus crocodilus (Lesuer, 1821)	6, 9, 13, 21, 24, 25, 30, 34, 37, 40	Occasional.	surface waters
Hemiramphidae			
Hemirhamphus far (Forsskål, 1775)	34, 37, 44	Occasional.	surface waters
Hyporhamphus dussumieri (Valenciennes, 1846)	8, 34	Two separate aggregations seen.	0–2
Zenarchopterus gilli Smith, 1945	4	Rare, but habitat (mangroves) not adequately sampled.	0–2
Holocentridae			
Myripristis adusta Bleeker, 1853	9, 10, 25, 45	Rare, about 6–7 individuals seen.	3–30
M. berndti Jordan & Evermann, 1902	7, 9, 10, 12, 14–16, 24–28, 33, 34, 37, 41, 45–47	Common.	8–55
M. hexagona (Lacepède, 1802)	2, 5, 20, 29	Occasional, but locally common at site 20.	10–40
M. kuntee Valenciennes, 1831	10, 12, 15, 26, 29, 33, 39, 45	Occasional.	5–30
M. murdjan (Forsskål, 1775)	13, 16, 27, 28, 30, 31, 32, 34	Occasional.	3–40
M. pralinia Cuvier, 1829	9, 12, 13, 15, 17, 24	Occasional, also collected with rotenone at one site.	3–30
M. violacea Bleeker, 1851	4, 8–11, 13–25, 29, 32, 33, 35, 42, 45–47	Common.	3–30
M. vittata Valenciennes, 1831	7, 9, 10, 12–16, 24, 47	Common on steep drop-offs.	12–80
Neoniphon argenteus (Valenciennes, 1831)	17–20	Rare, about five seen.	3–30
N. opercularis (Valenciennes, 1831)	25	Rare, only one seen.	3–30
N. sammara (Forsskål, 1775)	7, 12–15, 17–20, 25–28, 30, 34, 36	Occasional.	2–50
Sargocentron caudimaculatum (Rüppell, 1835)	7, 10, 12–17, 25–28, 30, 33, 34, 36–42, 45–47	Common.	6–45
S. diadema (Lacepède, 1802)	39	Rare, only one seen at depth of 15 m.	2–30
S. rubrum (Forsskål, 1775)	18–20	Rare.	
S. spiniferum (Forsskål, 1775)	8, 13, 14, 33–35, 41, 42	Occasional.	5–122
S. tiere (Cuvier, 1829)	42	One specimen collected with rotenone.	5–30
S. violaceus (Bleeker, 1853)	7, 24, 37, 45	Rare, only four seen.	3–30
Pegasidae			
Eurypegasus draconis (Linnaeus, 1766)	18	Rare, one pair seen.	3–12
Aulostomidae			
Aulostomus chinensis (Linnaeus, 1766)	1, 3, 5, 10, 13, 14, 25–27, 33, 34, 40, 44–46	Occasional.	2–122
Fistulariidae			
Fistularia commersoni Rüppell, 1835	1, 7, 12, 13, 21, 24, 31, 32, 37, 39, 46, 47	Occasional.	2–128

Species	Site Records	Abundance	Depth (m)
Centriscidae			
Aeoliscus strigatus (Günther, 1860)	29	Rare, several individuals seen. Normally common throughout most of Indonesia.	1–30
Syngnathidae			
Corythoichthys flavofasciatus (Rüppell, 1838)	3, 8, 18, 22, 31, 43	Occasional.	2–15
C. intestinalis (Ramsay, 1881)	4	Rare, one pair seen.	1–25
C. schultzi Herald, 1953	34	Rare, one seen.	
Doryrhamphus dactyliophorus (Bleeker, 1853)	15, 19, 31, 33, 40, 41	Occasional.	1–56
D. pessuliferus (Fowler, 1938)	40	Rare, only one seen.	5–30
Hippocampus kuda Bleeker, 1852	18	Rare, only one seen in weedy area.	1–10
Scorpaenidae			
Dendrochirus zebra (Cuvier, 1829)	36	Rare, only one seen.	3–25
Pterois antennata (Bloch, 1787)	2, 8, 13, 16, 36, 38, 41, 43	Occasional.	1–50
P. volitans (Linnaeus, 1758)	1, 3, 20, 27, 37, 38, 41, 42, 45, 46	Occasional.	2–50
Scorpaenodes albaiensis Evermann & Seale, 1907	13	One specimen collected with rotenone.	8–40
S. kelloggi (Jenkins, 1903)	43	A single specimen collected with rotenone.	1–25
S. parvipinnis (Garrett, 1863)	45	A single specimen collected with rotenone.	2–50
S. varipinnis Smith, 1957	13, 15	Two specimens collected with rotenone.	1–40
Scorpaenopsis oxycephala (Bleeker, 1849)	33, 43	Rare, only two seen, but very difficult to detect.	2–15
Sebastapistes cyanostigma (Bleeker, 1856)	38	Rarely seen, but a cryptic species that is seldom noticed.	2–15
Tetrarogidae			
Ablabys macracanthus (Bleeker, 1852)	29	Rare, three individuals seen together.	0–15
Synanceiidae			
Inimicus didactylus (Pallas, 1769)	43	Rare, only one seen, but difficult to detect due to excellent camouflage coloration.	1–40
Platycephalidae			
Cymbacephalus beauforti Knapp, 1973	20, 41	Rare, only two seen.	2–12
Thysanophrys arenicola Schultz, 1966	43	One specimen collected with rotenone.	1–80
T. chiltoni Schultz, 1966	37	One specimen collected with rotenone.	1–80

Serranidae

Species	Site Records	Abundance	Depth (m)
Aethaloperca rogaa (Forsskål, 1775)	12, 14, 15, 20, 27, 28, 33, 38, 40	Occasional.	1–55
Anyperodon leucogrammicus (Valenciennes, 1828)	10, 12–14, 19, 23, 24, 26, 38, 40, 47	Occasional.	5–50
Belonoperca chabanaudi Fowler and Bean, 1930			
Cephalopholis argus Bloch & Schneider, 1801	2, 3, 5, 11, 18, 19, 22, 23, 26, 42	Occasional.	1–45
C. boenack (Bloch, 1790)	2, 3, 5, 11, 18, 19, 22, 23, 26, 42	Occasional.	1–20
C. cyanostigma (Kuhl & Van Hasselt, 1828)	4–7, 10, 12–14, 17, 19, 20, 25–27, 30, 33, 38, 39, 43	Occasional.	2–35
C. leopardus (Lacepède, 1802)	1, 6, 7, 9, 12–14, 16, 17, 20, 21, 24, 25, 29, 37, 38	Occasional.	2–35
C. microprion (Bleeker, 1852)	4–6, 8, 10, 17, 19, 20, 22, 23, 29–31, 41, 43, 44	Occasional.	2–20
C. miniata (Forsskål, 1775)	12, 14, 16, 19, 20, 24, 25, 27, 33, 36, 37, 40, 42, 47	Occasional, except common at site 47.	3–150
C. polleni (Bleeker, 1868)	9, 12, 15, 16, 25, 47	Occasional, below 30 m depth on steep drop-offs.	25–120
C. sexmaculata Rüppell, 1828	9, 12–16, 19, 21, 24, 25, 47	Occasional under ledges on steep drop-offs; relatively common at site 47.	6–140
C. sonnerati (Valenciennes, 1828)	38	Rare, one seen at 30 m depth.	10–100
C. spiloparaea (Valenciennes, 1828)	10, 12, 13, 15, 16, 24, 25, 47	Occasional on steep drop-offs below 25 m depth.	16–108
C. urodeta (Schneider, 1801)	6, 7, 12–14, 25, 27, 28, 33, 34, 36–40, 42, 46, 47	Occasional.	1–36
Cromileptes altivelis (Valenciennes, 1828)	25–31	Rare, a large adult (20 m) and one juvenile seen.	2–40
Diploprion bifasciatum Cuvier, 1828	2–6, 24, 32, 33, 36, 41, 44	Occasional.	2–25
Epinephelus caeruleopunctatus (Bloch, 1790)	1	Rare, one adult seen.	5–25
E. corallicola (Kuhl & Van Hasselt, 1828)	3	Rare, one adult seen.	3–15
E. fasciatus (Forsskål, 1775)	12, 26–28, 33, 37, 42	Occasional.	4–160
E. hexagonatus (Bloch & Schneider, 1801)	37, 38, 41	Rare, only three seen.	3–10
E. macrospilos (Bleeker, 1855)	23	Rare, only one seen.	5–20
E. merra Bloch, 1793	6, 18, 25, 26, 29, 30, 33, 34, 37–39, 41	Occasional.	1–15
E. ongus (Bloch, 1790)	29, 32, 37	Rare, only three seen.	5–25
E. polyphekadion (Bleeker, 1849)	38, 43	Rare, two adults seen.	2–45
Gracila albimarginata (Fowler & Bean, 1930)	1, 25	Rare, about four individuals seen.	6–120
Grammistes sexlineatus (Thunberg, 1792)	36	Rare, only one seen.	3–30
Grammistinae sp.	14	Single juvenile collected with rotenone.	15–30
Liopropoma multilineatum Randall & Taylor, 1988	12	One specimen collected with rotenone.	11–50
Luzonichthys waitei (Fowler, 1931)	7, 13, 24, 27	Occasional, but locally abundant; usually found on steep drop-offs.	10–55
Plectropomus leopardus (Lacepède, 1802)	10, 13, 16, 43, 44	Occasional.	3–100
P. maculatus (Bloch, 1790)	1, 14, 20, 27, 38, 40	Occasional.	10–40
P. oligocanthus (Bleeker, 1854)	20, 21, 24, 25, 45	Occasional.	4–40
Pogonoperca punctata (Valenciennes, 1830)	46	Rare, an adult seen in 25 m depth.	20–150

Species	Site Records	Abundance	Depth (m)
Pseudanthias cooperi (Regan, 1902)	38	Rare, one small aggregation seen in 30 m depth.	15–60
P. dispar (Herre, 1955)	30, 47	Generally rare, but abundant at Site 47.	3–20
P. huchtii (Bleeker, 1857)	7, 9, 10, 12–17, 21, 24–27, 30–34, 39–42, 45	The most common Pseudanthias at both Togean and Banggai Islands.	4–20
P. hypselosoma Bleeker, 1878	1, 14, 20, 27, 38, 40	Occasional.	10–40
P. lori (Randall & Lubbock, 1981)	47	Generally rare except common at one site.	20–70
P. pleurotaenia (Bleeker, 1857)	7, 14, 16, 25, 27, 47	Occasional.	15–180
P. randalli (Lubbock and Allen, 1978)	12, 15, 25	Occasional, under ledges on steep drop-offs.	20–70
P. smithvanizi (Randall and Lubbock, 1981)	47	Generally rare, except common at one site.	6–70
P. squamipinnis (Peters, 1855)	1, 7, 26–28, 31, 33, 36–40, 46, 47	Common.	4–20
P. tuka (Herre & Montalban, 1927)	1, 5–7, 9, 10, 12–21, 24–27, 29, 30, 42, 45–47	Common.	8–25
Pseudogramma polyacantha (Bleeker, 1856)	13, 37	Two specimens collected with rotenone.	1–15
Variola albimarginata Baissac, 1953	13, 27, 33, 36, 37, 40, 42	Occasional.	12–90
V. Louti (Forsskål, 1775)	1, 8, 12, 25, 26, 30, 42	Occasional.	4–150
Cirrhitidae			
Amblycirrhitus bimacula (Jenkins, 1903)	25, 26	Rare.	1–10
Cirrhitichthys aprinus (Cuvier, 1829)	27, 39, 40	Rare.	5–40
C. falco Randall, 1963	1, 7,10, 12, 14, 16, 20, 24, 25, 32, 36, 42	Occasional.	4–45
C. oxycephalus (Bleeker, 1855)	13, 14, 37, 38, 42, 46, 47	Occasional.	2–40
Paracirrhites forsteri (Schneider, 1801)	1, 2, 7, 8–10, 12–14, 16, 20, 24, 25, 27, 28, 37–39, 46, 47	Moderately common.	1–35
Pseudochromidae			
Labracinus cyclophthalmus (Müller & Troschel, 1849)	29–42, 44–46	Common at Banggai Islands.	1–25
Pseudochromis bitaeniatus (Fowler, 1931)	2, 6, 10, 12, 15–22, 24, 39, 41, 42, 44, 45	Occasional.	5–30
P. elongatus Lubbock, 1980	13, 15, 19, 21, 25, 45	Occasional, also collected with rotenone.	8–25.
P. fuscus (Müller & Troschel, 1849)	Jan-47	Common, one of few species seen at every site.	1–30
P. marshallensis (Schultz, 1953)	40–42	Occasional.	2–25
P. paccagnellae Axelrod, 1973	1–3, 6, 6, 10, 12–21, 24–26, 37–39, 41, 42, 46, 47	Common.	6–70
P. perspicillatus Günther, 1862	31, 33, 40, 41, 44	Occasional at Banggai Islands.	3–20
P. polynema Fowler, 1931	1–3, 7, 9–21, 24, 25, 33, 38, 40–42, 44, 45, 47	Common.	10–30
Pseudoplesiops annae (Weber, 1913)	42	Several specimens collected with rotenone.	4–25
P. knighti Allen, 1987	15, 43, 45	Several specimens collected with rotenone.	5–35
P. sp.	43	Several specimens collected with rotenone.	5–35
P. typus Bleeker, 1858	37	One specimen collected with rotenone.	3–12

Species	Site Records	Abundance	Depth (m)
Plesiopidae			
Acanthoplesiops echinatus Smith–Vaniz & Johnson, 1990	42	One specimen collected with rotenone.	6–34
Calloplesiops altivelis (Steindachner, 1903)	33	Rare, but generally cryptic and difficult to detect.	3–45
Plesiops facicauus Mooi, 1995	42	One specimen collected with rotenone.	1–10
Opistognathidae			
Opistognathus sp. 1	31	Rare.	3–30
O. sp. 2	37	Rare.	0–10
O. sp. 3	1, 20, 40	Rare.	5–30
O. sp. 4	33	Rare.	10–30
Terapontidae			
Terapon jarbua (Forsskål, 1775)	3	Rare, but habitat (brackish waters and mangrove shores) not adequately sampled.	0–5
Priacanthidae			
Priacanthus hamrur (Forsskål, 1775)	9, 28, 39	Rare, only three seen.	5–80
Apogonidae			
Apogon angustatus (Smith & Radcliffe, 1911)	46	Rare, several seen at one site in 25 m depth.	5–30
A. aureus (Lacepède, 1802)	12, 13, 27, 30, 36, 38, 40	Occasional.	10–30
A. bandanensis Bleeker, 1854	2–5, 8, 17, 20	Occasional at Togean Islands.	5–20
A. caudicinctus Randall and Smith, 1988	15, 42	Several collected with rotenone.	0–12
A. ceramensis Bleeker, 1852	4	Generally rare, but locally common in mangroves at one site.	0–3
A. chrysopomus Bleeker, 1854	31, 32, 35	Generally rare, but locally common at three Banggai Islands sites.	1–18
A. chrysotaenia Bleeker, 1851	26–28, 32, 33, 36–38, 46	Occasional.	1–25
A. compressus (Smith & Radcliffe, 1911)	2–5, 15, 17, 18, 21, 22, 26, 29–31, 35, 42, 43, 45	Common.	2–20
A. crassiceps Garman, 1903	13, 15, 43, 47	Several specimens collected with rotenone.	1–25
A. cyanosoma Bleeker, 1853	12, 20, 30, 32, 36–44, 46, 47	Common, particularly at Banggai Islands.	3–22
A. dispar Fraser and Randall, 1976	14, 15, 19, 25, 47	Occasional, among black coral on steep drop–offs.	12–50
A. evermanni Jordan & Snyder, 1904	9, 15	Rare, but difficult to detect.	10–50
A. exostigma Jordan & Starks, 1906	6, 18, 43	Rare, about five individuals seen	3–25
A. fraenatus Valenciennes, 1832	3, 6, 12–14, 20, 23, 27, 38, 40, 41, 44	Occasional.	3–35
A. fragilis Smith, 1961	2, 3, 5, 11, 18, 19, 31, 43, 44	Occasional, but locally common.	1–15
A. franssedai Allen, Kuiter, and Randall, 1994	6	Rare.	15–40

Species	Site Records	Abundance	Depth (m)
A. fuscus Quoy & Gaimard, 1824	5, 20, 32, 40, 44	Occasional.	3–15
A. gilberti (Jordan and Seale, 1905)	2, 18, 19, 31, 44	Occasional, but locally common.	3–10
A. hartzfeldi Bleeker, 1852	29, 31, 37	Rare, only four seen.	1–10
A. hoeveni Bleeker, 1854	11, 18, 31, 32	Occasional, but locally abundant at Sites 31 and 32.	1–25
A. holotaenia Regan, 1905	36	Rare, except common around one coral head at Site 36.	10–30
A. kallopterus Bleeker, 1856	10, 13, 14, 20, 27, 33, 37, 38, 47	Occasional.	3–35
A. kiensis Jordan and Snyder, 1801	44	Rare, one individual seen on sand slope in 30 m depth.	12–30
A. leptacanthus Bleeker, 1856	4, 11, 31, 43	Occasional, but locally common.	1–12
A. moluccensis Valenciennes, 1832	33, 36, 41	Rare, a few individuals seen	3–35
A. multilineatus Bleeker, 1865	31, 37	Generally rare, but a secretive species that is difficult to detect during daylight hours.	1–5
A. neotes Allen, Kuiter, and Randall, 1994	2, 3–5, 8, 10, 14, 21, 22, 24, 33	Occasional.	10–25
A. nigrofasciatus Schultz, 1953	6, 9, 10, 12–17, 19–21, 24–27, 33, 36, 38, 39, 42, 45, 47	Common, but always in small numbers.	2–35
A. notatus (Houttuyn, 1782)	36	One aggregation of about 30 fish seen.	2–30
A. novemfasciatus Cuvier, 1828	37, 46	Rare.	0.5–3
A. ocellicaudus Allen, Kuiter, and Randall, 1994	3	Rare.	11–55
A. parvulus (Smith and Radcliffe, 1912)	31, 32, 39, 41–45	Occasional, but locally common.	4–20
A sealei Fowler, 1918	6, 8, 15, 20, 29, 31, 37, 41, 43, 44	Common.	1–15
A. selas Randall and Hayashi, 1990	2, 5, 43	Occasional in sheltered situations	20–35
A. thermalis Cuvier, 1829	4, 11	Rare, two aggregations seen.	0–20
A. timorensis Bleeker, 1873	37	Collected with rotenone at one site.	0–3
A. trimaculatus Cuvier, 1828	2, 5, 17, 37	Occasional.	2–10
A. wassinki Bleeker, 1860	3, 15, 26, 27, 31	Occasional.	2–30
Apogonichthys perdix Bleeker, 1854	37	Collected with rotenone.	1–65
Archamia fucata (Cantor, 1850)	2–6, 15, 19, 20, 24, 26, 31, 43	Occasional.	3–60
A. zosterophora (Bleeker, 1858)	2, 4, 5, 8, 11, 15, 17–20, 31, 43, 44	Occasional.	2–15
Cheilodipterus alleni Gon, 1993	4, 5, 9, 17, 19	Rare, only five seen.	1–25
C. artus Smith, 1961	5, 7, 9, 10, 12, 13, 15, 16, 23, 25, 26, 29, 31, 38, 41, 42, 46, 47	Occasional.	3–15
C. macrodon Lacepède, 1801	1–3, 5, 6, 11, 14, 15, 18, 19, 26, 41, 42, 45	Occasional.	4–30
C. nigrotaeniatus Smith and Radcliffe, 1912	2–5, 8, 11, 31, 35, 44	Occasional.	2–25
C. quinquelineatus Cuvier, 1828	1–6, 8–47	Common, seen at nearly every site.	1–40
Fowleria aurita (Valenciennes, 1831)	14, 43	Several specimens collected with rotenone.	0–15
F. variegata (Valenciennes, 1832)	37	One specimen collected with rotenone.	
Gymnapogon sp.	17	One specimen collected with rotenone.	5–25

Species	Site Records	Abundance	Depth (m)
Pseudamia gelatinosa Smith, 1955	43	One specimen collected with rotenone.	1–40
Pterapogon kauderni Koumans, 1933	31, 32, 35, 43, 44	Locally common in sheltered locations of larger islands in the Banggai Group.	1–2
Rhabdamia cypselurus Weber, 1909	6	Rare, only one aggregation seen	2–15
R. gracilis (Bleeker, 1856)	3, 36, 38, 40, 41, 45	Occasional.	5–20
Siphamia jebbi Allen, 1993	43	Several specimens collected with rotenone.	14–30
S. majimae Matsubara & Iwai, 1958	36	Four specimens collected from *Diadema* urchin.	5–25
Sphaeramia nematoptera (Bleeker, 1856)	2–5, 11, 17–20, 22, 24, 31, 35, 43, 44	Occasional.	1–8
S. orbicularis (Cuvier, 1828)	3, 4, 20, 32	Occasional, but its primary habitat (mangroves) not properly sampled.	0–3
Malacanthidae			
Hoplolatilus starcki Randall & Dooley, 1974	7, 14, 16	Generally rare, but common at Site 16.	20–105
Malacanthus brevirostris Guichenot, 1848	6, 27, 38	Rare.	10–45
M. latovittatus (Lacepède, 1798)	6, 25, 27, 36, 40, 42	Occasional.	5–30
Carangidae			
Alepes sp.	41	Generally rare, several seen at one site.	0–20
Carangoides bajad (Forsskål, 1775)	7, 17, 19, 21, 47	Occasional.	5–30
C. ferdau (Forsskål, 1775)	41	Rare, only one seen.	2–40
C. fulvoguttatus (Forsskål, 1775)	15, 16, 24, 38, 40, 42, 44, 46, 47	Occasional.	5–30
C. plagiotaenia Bleeker, 1857	1–5, 7, 9, 10, 12, 17, 19, 21, 22, 24, 25, 37, 43–45, 47	Occasional, the most common carangid seen.	5–200
Caranx melampygus Cuvier, 1833	1, 5–8, 12–16, 28, 45	Occasional.	1–190
C. papuensis Alleyne and Macleay, 1877	2, 31	Rare, only two seen.	2–30
C. sexfasciatus Quoy & Gaimard, 1825	14	Rare.	3–96
Elegatis bipinnulatus (Quoy & Gaimard, 1825)	13, 39, 43	Occasional.	5–150
Gnathanodon speciosus (Forsskål, 1775)	4	Rare, one juvenile seen.	1–30
Scomberoides lysan (Forsskål, 1775)	26, 33	Rare, only two seen.	1–100
Selaroides leptolepis (Kuhl & van Hasselt, 1833)	31	A large school seen at Site 31	1–15
Lutjanidae			
Aphareus furca (Lacepède, 1802)	1, 12–14, 17, 24, 25	Occasional.	6–70
Aprion virescens Valenciennes, 1830	26, 31, 40, 42, 45	Occasional, several large individuals seen at Site 42.	2–100
Lutjanus bengalensis (Bloch, 1790)	1	Rare, one juvenile seen.	10–25
L. biguttatus (Valenciennes, 1830)	3–5, 9, 10, 17, 19, 21, 23–27, 29, 45	Moderately common.	3–40
L. bohar (Forsskål, 1775)	1, 5–10, 12–15, 18, 21, 23–30, 33, 34, 37–42, 45, 47	Common.	4–180

Species	Site Records	Abundance	Depth (m)
L. boutton (Lacepède, 1802)	13, 29	Rare, only two seen.	5–25
L. carponotatus (Richardson, 1842)	2–4, 6, 8, 11, 18–20, 22, 23, 41, 43–45	Moderately common.	2–35
L. decussatus (Cuvier, 1828)	1–30, 38, 41, 44–46	Common.	5–30
L. ehrenburgi (Peters, 1869)	4, 18, 20	Rare, a few seen on the edge of mangroves.	1–20
L. fulviflamma (Forsskål, 1775)	3, 4, 25, 26, 28, 40, 46	Occasional.	1–35
L. fulvus (Schneider, 1801)	3, 4, 12–14, 18, 20, 28, 30	Occasional.	2–40
L. gibbus (Forsskål, 1775)	1, 12–15, 18, 24, 26, 28, 30, 37, 44	Occasional.	6–40
L. kasmira (Forsskål, 1775)	10, 12, 14, 15, 17, 20, 25	Occasional.	3–265
L. lunulatus (Park, 1797)	27	Rare, one seen in 30 m depth.	10–40
L. lutjanus Bloch, 1790	26, 28, 31	Occasional.	4–30
L. monostigma (Cuvier, 1828)	13, 15, 20, 21, 25, 27, 37, 39, 47	Occasional.	5–60
L. quinquelineatus (Bloch, 1790)	39	Rare, one seen in 15 m depth.	5–30
L. rivulatus (Cuvier, 1828)	2, 10, 12, 13	Rare, four individuals seen.	2–100
L. rufolineatus (Valenciennes, 1830)	3, 14, 26	Occasional aggregations sighted.	12–50
L. vitta (Quoy & Gaimard, 1824)	31	Rare, a few small individuals encountered	8–40
Macolor macularis Fowler, 1931	1, 3, 5, 6, 8–16, 19, 20, 24–27, 29, 30, 37–39, 42, 45, 47	Common.	3–50
M. niger (Forsskål, 1775)	10, 38, 45	Occasional.	3–90
Paracaesio sordidus Abe and Shinohara, 1962	12	Rare, one aggregation encountered on steep drop–off.	30–200
Symphorus nematophorus (Bleeker, 1860)	3, 41, 42	Rare, three individuals seen.	5–50
Caesionidae			
Caesio caerulaurea Lacepède, 1802	11, 20, 23	Occasional schools encountered	1–30
C. cuning (Bloch, 1791)	1–5, 7–11, 15, 16, 18–32, 35, 39, 40, 43, 44, 47	Common.	1–30
C. lunaris Cuvier, 1830	6, 12, 13, 21, 25, 47	Occasional.	1–35
C. teres Seale, 1906	7, 12–14, 27, 33, 39, 40, 42	Occasional.	1–40
Dipterygonotus balteatus (Valenciennes, 1830)	15, 20, 29	Rare, but probably inadequately sampled as it mixes with Pterocaesio and is difficult to detect.	1–25
Pterocaesio digramma (Bleeker, 1865)	5, 13	Two schools encountered.	1–25
P. marri Schultz, 1953	10, 27, 29, 32, 33, 37–39, 45	Occasional.	1–35
P. pisang (Bleeker, 1853)	5, 6, 9, 15, 17, 19–21, 23, 25, 27, 47	Moderately common.	1–35
P. randalli Carpenter, 1987	9, 12, 14–16, 21, 25, 47	Occasional, except abundant on steep drop–offs at Site 47.	10–40
P. tessellata Carpenter, 1987	25	Rare, one aggregation seen.	1–35
P. tile (Cuvier, 1830)	7, 12, 14, 16, 25–28, 30, 33, 38, 47	Moderately common.	1–60

Species	Site Records	Abundance	Depth (m)
Gerreidae			
Gerres abbreviatus Bleeker, 1850	29	Rare.	0–40
G. argyreus (Schneider, 1801)	3, 18, 31	Occasional.	0–5
G. oyena (Forsskål, 1775)	30, 34, 37	Occasional.	0–10
Haemulidae			
Diagramma sp.	31	Seen at only one site, but common there.	3–30
Plectorhinchus chaetodontoides (Lacepède, 1800)	9, 10, 13–15, 17, 21, 24, 25, 29, 31, 35, 43, 45–47	Moderately common	1–40
P. gibbosus (Lacepède, 1802)	16	Rare, one large individual sighted in 55 m depth	2–30
P. lessoni (Cuvier, 1830)	2, 6, 9, 10, 16, 20, 23, 32, 36, 40, 41, 46	Moderately common.	5–35
P. lineatus (Linnaeus, 1758)	10, 13, 33, 37, 40, 45, 47	Occasional, except common at Site 13.	2–40
P. orientalis (Bloch, 1793)	7, 26–28, 32–34, 36–37	Occasional.	3–30
P. polytaenia (Bleeker, 1852)	27, 33, 40	Rare, only three seen.	3–30
Lethrinidae			
Gnathodentex aurolineatus Lacepède, 1802	26, 45	Rare, several individuals seen at two sites.	1–30
Lethrinus atkinsoni Seale, 1909	41	Rare, two seen in 3 m depth.	1–15
L. erythracanthus Valenciennes, 1830	5, 9, 10, 13, 17, 23, 25	Occasional, mainly juveniles sighted, except large adult seen at Site 9.	15–120
L. erythropterus Valenciennes, 1830	1–11, 16, 18–25, 42	Moderately common, at least at the Togean Islands.	2–30
L. harak (Forsskkal, 1775)	8, 18, 20, 25, 29–35, 37	Moderately common, at least at the Banggai Islands.	1–20
L. obsoletus (Forsskål, 1775)	13, 18, 20, 25, 27	Occasional.	5–15
L. olivaceous Valenciennes, 1830	13, 17, 25, 26, 28	Occasional.	4–185
L. ornatus Valenciennes, 1830	17, 20, 37, 43, 46	Occasional.	3–20
L. variegatus Valenciennes, 1830	13	Rare.	1–10
L. xanthocheilus Klunzinger, 1870	20, 26, 37	Rare, several large individuals seen in the shallows.	2–25
Monotaxis grandoculis (Forsskål, 1775)	3–10, 12–30, 32, 33, 35–40, 42, 45–47	Common.	1–100
Nemipteridae			
Pentapodus sp.	2, 3, 6, 10, 17, 18, 22, 23, 26, 27, 30–32, 34–46	Common.	3–25
P. trivittatus (Bloch, 1791)	2–6, 8–10, 18, 22, 23, 31, 33, 35–37, 41, 43, 44	Moderately common	1–15
Scolopsis affinis Peters, 1876	3, 11, 14, 18, 20, 26–28, 31, 34, 40	Moderately common.	3–60
S. bilineatus (Bloch, 1793)	7, 9, 10, 12, 13, 15, 16, 20, 25–42, 44, 46, 47	Common.	2–20
S. ciliatus (Lacepède, 1802)	4, 11, 20, 29	Occasional in sheltered areas with reduced visibility.	1–30

Species	Site Records	Abundance	Depth (m)
S. lineatus Quoy & Gaimard, 1824	17–20, 30, 32, 33, 35, 44–47	Moderately common.	0–10
S. margaritifer (Cuvier, 1830)	1–7, 9, 10, 15–23, 25–29, 31, 35, 38–41, 43, 44	Common.	2–20
S. monogramma (Kuhl & Van Hasselt, 1830)	8, 18, 31, 43	Occasional.	5–50
S. trilineatus Kner, 1868	30	Rare, a few seen in shallow water.	1–10
S. xenochrous Günther, 1792)	26, 27, 32–36	Occasional, except common at Site 27.	5–50
Mullidae			
Mulloidichthys flavolineatus (Lacepède, 1802)	6, 8, 9, 16, 19, 20, 25, 37, 41, 47	Occasional.	1–40
M. vanicolensis (Valenciennes, 1831)	28, 36, 43	Occasional.	1–113
Parupeneus barberinus (Lacepède, 1801)	1, 2, 4–10, 13–47	Common.	1–100
P. bifasciatus (Lacepède, 1801)	1, 6–10, 12–28, 30, 33–42, 45–47	Common.	1–80
P. cyclostomus (Lacepède, 1802)	1, 6, 10, 13, 14, 24, 25, 27, 28, 33, 36, 39–41, 44, 46, 47	Occasional.	2–92
P. indicus (Shaw, 1803)	13, 20	Rare, only two seen.	1–20
P. multifasciatus Bleeker, 1873	1, 3–10, 12–47	Common.	1–140
P. pleurostigma (Bennett, 1830)	26, 27, 36	Rare.	5–46
Upeneus tragula Richardson, 1846	4, 8, 29, 31, 37, 41	Occasional.	0–30
Pempheridae			
Parapriacanthus ransonneti Steindachner, 1870	6, 15, 20, 33, 40, 42, 46	Occasional.	5–30
Pempheris oualensis Cuvier, 1831	9, 10, 13, 29, 32, 33	Occasional, except huge shoals encountered at Sites 32 and 33.	3–38
P. vanicolensis Cuvier, 1831	28, 29	Only two aggregations encountered, but a secretive species that shelters in caves during the day.	2–35
Toxotidae			
Toxotes jaculatrix (Pallas, 1767)	3, 4, 20	Rare, but habitat (mangroves) inadequately surveyed.	0–2
Kyphosidae			
Kyphosus bigibbus Lacepède, 1801	13, 25	Rare.	1–25
K. cinerascens (Forsskål, 1775)	3, 7, 13, 47	Occasional.	1–24
K. vaigiensis (Quoy & Gaimard, 1825)	12, 14, 47	Occasional.	1–20
Monodactylidae			
Monodactylus argenteus (Linnaeus, 1758)	4, 8	Rare, but habitat inadequately surveyed	0–5

Species	Site Records	Abundance	Depth (m)
Chaetodontidae			
Chaetodon adiergastos Seale, 1910	30, 32, 41, 43	Rare, only five seen.	1–25
C. auriga Forsskål, 1775	1, 13, 27, 30, 32, 34, 35, 39, 41, 43	Occasional.	1–30
C. baronessa Cuvier, 1831	1, 3, 5–8, 10, 12–30, 32–34, 36, 37, 40–42, 44–47	Common.	2–15
C. bennetti Cuvier, 1831	7, 9, 10, 13, 15–17, 26, 31, 35, 37, 38	Occasional.	5–30
C. burgessi Allen and Starck, 1973	16, 25	Rare, only two pairs sighted in 22–30 m depth.	30–80
C. citrinellus Cuvier, 1831	6, 13, 14, 28–30, 33, 34, 37, 40, 42, 46, 47	Occasional.	1–12
C. ephippium Cuvier, 1831	6, 7, 9, 10, 12, 13, 16, 18–20, 24, 25–28, 41, 45	Occasional.	1–30
C. guentheri Ahl, 1913	27	Rare, one pair observed in 30 m depth.	5–30
C. kleinii Bloch, 1790	1, 3, 5–10, 12–16, 18–21, 24–42, 44–47	Common.	6–60
C. lineolatus Cuvier, 1831	28, 30, 32, 35	Occasional.	2–170
C. lunula Lacepède, 1803	6, 8, 10, 13–15, 19,21, 25, 28, 30, 32, 33, 41, 47	Occasional.	1–40
C. lunulatus Quoy and Gaimard, 1824	Jan–47	The most common butterflyfish and one of few fishes seen at every site.	1–25
C. melannotus Schneider, 1801	13, 26, 28, 30, 32–34, 36, 39, 40, 46	Occasional.	2–15
C. meyeri Schneider, 1801	28, 29	Rare, only two seen.	5–25
C. ocellicaudus Cuvier, 1831	5, 7, 11, 13, 15, 20, 22, 25, 26, 29, 32–34, 40, 43–45	Occasional.	1–15
C. octofasciatus Bloch, 1787	2–5, 8, 11, 19–23, 31, 43, 44	Moderately common.	3–20
C. ornatissimus Cuvier, 1831	1, 7–10, 12–25, 27, 28, 42, 46	Moderately common, at least in the Togean Islands.	1–36
C. oxycephalus Bleeker, 1853	5, 14–16, 25, 27, 31–33, 35, 37	Occasional.	8–30
C. punctatofasciatus Cuvier, 1831	1, 6, 7, 9, 10, 12, 13, 15–17, 19–21, 24–26, 28–30, 32–34, 37–40, 42, 45–47	Moderately common.	6–45
C. rafflesi Bennett, 1830	5, 7–13, 16–30, 32, 41, 44	Moderately common.	1–15
C. selene Bleeker, 1853	26–28, 30–32, 36, 40, 41	Not seen in Togean Islands, but occasional at mainland and at the Banggai Islands.	15–40
C. semeion Bleeker, 1855	7, 8, 10, 12, 13, 15, 19–21, 24, 25, 34, 35, 46	Occasional.	1–25
C. speculum Cuvier, 1831	12, 30, 31, 34, 42, 44	Occasional.	3–20
C. trifascialis Quoy & Gaimard, 1824	1, 5, 7, 9, 10, 12, 13, 15–17, 21, 24–30, 32–34, 36–38, 40, 43–47	Common.	2–30
C. ulietensis Cuvier, 1831	1, 6, 7, 10, 12–22, 24–30, 32–38, 45–47	Moderately common.	8–30
C. unimaculatus Bloch, 1787	12, 13, 16, 25–28, 30, 34, 37–39	Occasional.	5–60
C. vagabundus Linnaeus, 1758	1, 3, 5–10, 12–23, 25–30, 33, 34, 36–41, 44, 47	Common.	1–30
C. xanthurus Bleeker, 1857	19, 29, 37, 41, 42, 46	Occasional.	10–40
Chelmon rostratus (Linnaeus, 1758)	44	Rare, one pair encountered in 14 m depth.	1–15

Species	Site Records	Abundance	Depth (m)
Coradion chrysozonus Cuvier, 1831	1–4, 6, 7, 10, 13, 16, 17, 20–23, 25–27, 31–41, 43, 44	Moderately common to a depth of 33 m.	5–60
Forcipiger flavissimus Jordan & McGregor, 1898	10, 27, 29, 33, 34, 37–39, 41, 42, 46	Occasional, but mainly seen at Banggai Islands.	2–114
F. longirostris (Broussonet, 1782	6, 7, 9, 10, 12–17, 19–21, 24, 25, 29, 37, 45–47	Moderately common.	5–60
Hemitaurichthys polylepis (Bleeker, 1857)	7, 9, 13–16, 24, 27, 37, 46, 47	Occasional, usually on steep drop-offs.	3–40
Heniochus acuminatus (Linnaeus, 1758)	9, 20, 36, 47	Rare, only four seen.	2–75
H. chrysostomus Cuvier, 1831	1, 4, 6–10, 12–18, 21, 24, 25, 28–30, 35–37, 44–47	Moderately common.	5–40
H. diphreutes Jordan, 1903	14, 16, 27	Occasional, to depths of over 60 m; super-abundant at Site 27.	15–210
H. singularius Smith & Radcliffe, 1911	1, 2, 4, 7, 9, 10, 12–17, 21, 24, 25, 32	Occasional.	12–45
H. varius (Cuvier, 1829)	1, 2, 5–47	Common.	2–30
Pomacanthidae			
Apolemichthys trimaculatus (Lacepède, 1831)	10, 16, 25, 28, 46	Occasional.	10–50
Centropyge bicolor (Bloch, 1798)	1, 6, 10, 13, 14, 18, 20, 24–30, 32–34, 36–42, 46, 47	Moderately common.	3–35
C. bispinosus (Günther, 1860)	1, 7, 10, 12, 16, 17, 24–27	Occasional.	10–45
C. flavicauda Fraser-Brunner, 1933	34, 36, 46	Occasional, but locally common in rubble between 10–15 m depth at Sites 34 and 36.	10–60
C. multifasciatus (Smith & Radcliffe, 1911)	1, 7, 9, 10, 12, 14–17, 20, 24, 25, 42, 45, 47	Occasional.	10–50
C. nox (Bleeker, 1853)	1, 5, 7, 9, 10, 12–15, 17, 20, 25, 26, 39, 42, 45	Occasional.	10–70
C. tibicen (Cuvier, 1831)	3, 10, 14, 25–43, 45, 47	Moderately common.	3–40
C. vroliki (Bleeker, 1853)	1, 6–10, 12–30, 34, 36–47	Common.	3–25
Chaetodontoplus mesoleucus (Bloch, 1787)	1–6, 8, 10, 17–20, 22, 23, 31–35, 41, 43–45	Moderately common.	1–20
Genicanthus lamarck Lacepède, 1798	29, 33, 34, 36	Occasional.	15–40
Pomacanthus imperator (Bloch, 1787)	1, 10, 12–14, 16, 24–28, 33, 40–42, 45–47	Occasional.	3–70
P. navarchus Cuvier, 1831	1, 13, 24, 29, 33, 37, 38, 42, 43, 45, 46	Occasional.	3–30
P. sexstriatus Cuvier, 1831	4, 6, 13, 16, 23, 27, 31, 35, 42–45	Occasional, to 50 m depth.	3–50
P. xanthometopon (Bleeker, 1853)	9, 10, 13–15, 21, 24, 47	Occasional.	5–30
Pygoplites diacanthus (Boddaert, 1772)	1–3, 5, 10, 12–29, 33, 35, 37–42, 44–47	Common.	3–50
Mugilidae			
Liza vaigiensis (Quoy & Gaimard, 1825)	3, 29, 37	Occasional schools encountered	0–3

Pomacentridae

Species	Site Records	Abundance	Depth (m)
Abudefduf lorenzi Hensley and Allen, 1977	3, 8, 20, 29, 32	Occasional.	0–6
A. sexfasciatus Lacepède, 1802	18, 20, 25, 29, 31, 33, 34, 36, 38, 40, 41	Occasional, excepth common at Site 33.	1–15
A. vaigiensis (Quoy & Gaimard, 1825)	4, 8, 13, 14, 18, 21, 25, 27–36, 40, 42, 46, 47	Moderately common.	1–12
Acanthochromis polyacantha (Bleeker, 1855)	30–36, 40, 43	Occasional, but locally abundant at Site 30; seen only at the Banggai Islands.	1–5
Amblyglyphidodon aureus (Cuvier, 1830)	1–3, 6, 7, 9, 10, 12–16, 19–21, 24, 25, 27, 32, 37–42, 46, 47	Moderately common.	10–35
A. batunai Allen, 1995	2–4, 11, 19, 22, 23, 30–33, 35, 44, 45	Moderately common.	2–12
A. curacao (Bloch, 1787)	2–13, 15–25, 29–47	Common.	1–15
A. leucogaster (Bleeker, 1847)	1–10, 15, 17, 19–21, 24–34, 37–47	Common.	2–45
Amblypomacentrus breviceps (Schlegel and Müller, 1839)	31, 44	Rare, several seen on sand slopes at depths between 16–25 m	2–35
A. clarus Allen and Adrim, 2000	31	An undescribed species seen only in Banggai Harbour at depth of 16 m.	12–30
Amphiprion clarkii (Bennett, 1830)	1, 2, 5–8, 10, 12–16, 18–21, 23–33, 35–47	Moderately common.	1–55
A. melanopus Bleeker, 1852	29, 30, 32, 33, 35, 43, 46	Occasional, except relatively abundant at Site 30.	1–10
A. ocellaris (Cuvier, 1830)	3, 4, 12, 14, 15, 22–25, 29, 32, 33, 35, 40, 47	Occasional.	1–15
A. perideraion Bleeker, 1855	9, 12, 13, 15, 16, 20, 24, 27, 29, 33, 35, 36, 40, 41, 43, 45	Occasional.	3–20
A. polymnus (Linnaeus, 1758)	31, 32	Rare, two groups encountered at depths of 15–35 m	2–35
A. sandaracinos Allen, 1972	12, 23, 30, 33, 40, 42, 45	Occasional.	3–20
Cheiloprion labiatus (Day, 1877)	30, 32–35	Occasional.	1–3
Chromis alpha Randall, 1988	7, 9, 10, 12–17, 19, 24, 25, 38, 47	Moderately common on steep outer reef slopes.	18–95
C. amboinensis (Bleeker, 1873)	1–3, 6, 7, 9, 10, 12–17, 19–21, 24, 25, 29, 30, 37–42, 45–47	Common.	5–65
C. analis (Cuvier, 1830)	7, 9, 10, 12–14, 16, 24, 25, 37, 42, 47	Moderately common, but locally abundant at Sites 16 and 25.	10–70
C. atripectoralis Welander & Schultz, 1951	1, 9, 13, 25, 29, 30, 32–34, 36–40, 42, 45–47	Moderately common.	2–15
C. atripes Fowler and Bean, 1928	1, 7, 9, 10, 12–16, 19, 20, 25, 29, 30, 37–39, 42, 45–47	Moderately common.	10–35
C. caudalis Randall, 1988	9, 10, 13, 14, 16, 21, 25, 27, 30, 33, 34, 37–39, 47	Moderately common on steep outer reef slopes.	20–50
C. delta Randall, 1988	1, 6, 7, 9, 12–16, 21, 24, 25, 27, 29, 30, 37, 38, 42, 46, 47	Moderately common	10–80
C. elerae Fowler & Bean, 1928	3, 9, 12–16, 19, 21, 25, 47	Occasional, but locally common on steep outer reef slopes.	12–70
C. lepidolepis Bleeker, 1877	1, 3, 6, 7, 9, 10, 12–14, 21, 24–30, 32, 36, 38, 40–42, 45–47	Common.	2–20
C. lineata Fowler & Bean, 1928	1, 7, 9, 12–16, 19, 21, 24, 25, 34, 46, 47	Moderately common.	2–10

Species	Site Records	Abundance	Depth (m)
C. margaritifer Fowler, 1946	7, 9, 10, 12–16, 21, 25–30, 32–34, 36–42, 46, 47	Moderately common.	2–20
C. retrofasciata Weber, 1913	1, 2, 6, 7, 9, 10, 12–17, 19–25, 27, 29, 30, 37, 39, 42, 47	Moderately common.	5–65
C. scotochiloptera Fowler, 1918	27, 28, 33, 40, 47	Occasional, except common at Site 27.	5–20
C. ternatensis (Bleeker, 1856)	1–10, 12–32, 36–40, 42, 45–47	Abundant	2–15
C. viridis (Cuvier, 1830)	3, 5, 7, 12, 13, 17, 19–21, 24, 25, 27, 29, 30, 33–44	Common.	1–12
C. weberi Fowler & Bean, 1928	6, 9, 10, 13, 14, 18, 20, 21, 25–28, 30–34, 36–40, 42, 45–47	Common.	3–25
C. xanthochira (Bleeker, 1851)	37, 38, 42, 47	Occasional.	10–48
C. xanthura (Bleeker, 1854)	1–3, 6–10, 12–16, 19–21, 23–28, 30, 34, 36–39, 41, 42, 45–47	Common.	2–30
Chrysiptera biocellata (Quoy & Gaimard, 1824)	18, 20, 33	Seen only on three occasions, but probably common on shallow reefs.	0–5
C. bleekeri (Fowler and Bean, 1928)	30–34, 36, 39	Occasional, but locally common at Sites 30 and 34.	3–15
C. brownriggii (Bennett, 1828)	28	Rare, only three seen.	0–2
C. caruleolineata (Allen, 1973)	16	Rare, except locally common at one site at depths of 35 m	30–65
C. cyanea (Quoy & Gaimard, 1824)	3, 6, 8, 16–18, 20, 21, 29, 30, 32–34, 37, 43, 44	Occasional.	1–12
C. oxycephala (Bleeker, 1877)	2–5, 11, 18, 19, 22, 23, 35, 43, 44	Moderately common.	1–16
C. parasema (Fowler, 1918)	2, 4, 11, 35, 43, 44	Occasional, in sheltered areas with good coral cover but reduced visibility.	1–16
C. rex (Snyder, 1909)	1, 6–10, 13–21, 23–26, 28, 29, 34, 36–38, 40, 41, 46–47	Moderately common.	1–6
C. rollandi (Whitley, 1961)	1, 3–10, 12–27, 29–35, 37–44, 46, 47	Common.	2–35
C. springeri Allen and Lubbock, 1976	1–13, 15–25, 29, 31, 32, 35, 41–45	Common.	5–30
C. talboti (Allen, 1975)	10, 25, 27, 30, 33, 37–40, 42, 46, 47	Occasional.	6–35
C. unimaculata (Cuvier, 1830)	1, 6, 16–18, 25, 30–33, 41	Occasional.	0–2
Dascyllus aruanus (Linnaeus, 1758)	2–6, 8, 11, 15, 17, 19, 21, 25, 29–31, 33, 35, 37, 38, 41–44	Moderately common.	1–12
D. melanurus Bleeker, 1854	1–5, 8, 11, 17–23, 25, 29–33, 35, 43, 44	Moderately common.	1–10
D. reticulatus (Richardson, 1846)	1, 6–10, 12–16, 20, 21, 23–28, 30–34, 36–42, 44, 46, 47	Common.	1–50
D. trimaculatus (Rüppell, 1928)	3, 4–7, 9, 10, 12–16, 18–20, 33–47	Common.	1–55
Dischistodus chrysopoecilus (Schlegel & Müller, 1839)	4, 18, 20, 29, 30, 32, 35	Occasional.	1–5
D. melanotus (Bleeker, 1858)	2–6, 8, 9, 15, 17, 18, 20–23, 25, 29–31, 33, 39	Occasional.	1–10
D. perspicillatus (Cuvier, 1830)	2–5, 8, 11, 15, 16, 18–20, 22, 23, 25, 29–31, 35, 43, 44	Occasional.	1–10
D. prosopotaenia (Bleeker, 1852)	29, 31, 35, 43–45	Occasional.	1–12
D. pseudochrysopoecilus Allen and Robertson, 1974	30, 33, 41	Rare.	1–5
Hemiglyphidodon plagiometopon (Bleeker, 1852)	2–5, 8, 11, 17–20, 22, 23, 29, 35, 43, 44	Occasional.	1–20
Lepidozygus tapeinosoma (Bleeker, 1856)	7, 12, 25, 29, 30	Occasional.	5–25

Species	Site Records	Abundance	Depth (m)
Neoglyphidodon crossi Allen, 1991	13, 25–30, 34, 36, 40, 46	Moderately common.	2–5
N. melas (Cuvier, 1830)	1–10, 12, 17, 18, 20–30, 32–34, 36, 38–47	Common.	1–12
N. nigroris (Cuvier, 1830)	1, 5–10, 13–17, 19, 21, 24–26, 28–30, 32–42, 45–47	Common.	2–23
N. oxyodon (Bleeker, 1857)	30, 32–35	Occasional, but locally common at Site 30.	0–4
N. thoracotaeniatus (Fowler and Bean, 1928)	5, 21, 24, 25, 29, 45	Occasional.	15–45
Neopomacentrus cyanomos (Bleeker, 1856)	1, 31, 33, 40, 41, 44, 45	Occasional.	5–18
N. filamentosus (Macleay, 1833)	2, 3, 8, 11, 18	Occasional.	5–12
N. nemurus (Bleeker, 1857)	2–4, 23, 29, 43	Occasional.	1–10
N. violascens (Bleeker, 1848)	31	Rare, one seen in 20 m depth.	1–25
Plectroglyphidodon dickii (Lienard, 1839)	7, 9, 13, 25, 26, 28, 33, 34, 36, 38, 40, 46, 47	Occasional.	1–12
P. lacrymatus (Quoy & Gaimard, 1824)	1–10, 12–30, 32–47	Common.	2–12
P. leucozonus (Bleeker, 1859)	8, 28	Rare.	0–2
Pomacentrus adelus Allen, 1991	2–6, 8, 16–23, 25, 29, 30, 32–44, 47	Common.	0–8
P. alexanderae Evermann & Seale, 1907	1–6, 8, 10, 11, 15, 17–24, 29, 31, 32, 35, 41, 43–45	Common.	5–30
P. amboinensis Bleeker, 1868	1–3, 6–10, 13–17, 19–21, 23, 25–30, 32–47	Common.	2–40
P. auriventris Allen, 1991	12–14, 16, 25–28, 30–34, 36–40, 42, 43, 46, 47	Moderately common.	2–15
P. bankanensis Bleeker, 1853	1, 6, 8–10, 12–14, 16–21, 24–30, 33, 34, 36–42, 45–47	Common.	0–12
P. brachialis Cuvier, 1830	1, 6, 7, 9, 10, 12–16, 19, 20, 24–30, 32–34, 36–42, 45–47	Common.	6–40
P. burroughi Fowler, 1918	2–5, 8, 11, 15, 17–23, 35, 42–44	Moderately common.	2–16
P. chrysurus Cuvier, 1830	10, 39	Rare.	0–3
P. coelestis Jordan & Starks, 1901	1, 6, 7, 9, 10, 12–14, 16, 19–21, 24–28, 30, 32–34, 36–42, 46, 47	Common.	1–12
P. cuneatus Allen, 1991	3, 11, 18	Generally rare, but several seen at three sites.	1–6
P. grammorhynchus Fowler, 1918	30, 32, 35, 43	Occasional.	2–12
P. lepidogenys Fowler & Bean, 1928	1, 13, 15, 21, 25–30, 33, 36, 38–40, 45–47	Moderately common.	1–12
P. moluccensis Bleeker, 1853	1–3, 5–10, 12–22, 24–47	Common.	1–14
P. nagasakiensis Tanaka, 1917	3, 6, 10, 20, 22, 27–30, 32–34, 36–44	Moderately common.	5–30
P. nigromarginatus Allen, 1973	1, 2, 5, 7, 9, 10, 12–17, 19–21, 23–25, 38, 39, 42, 46, 47	Moderately common.	20–50
P. opisthostigma Fowler, 1918	2, 11, 18, 31, 35	Occasional.	6–15
P. pavo (Bloch, 1878)	29, 31	Rare.	1–16
P. philippinus Evermann & Seale, 1907	2–6, 8–10, 13, 15, 17–25, 27, 28, 38, 40, 41, 42, 46, 47	Moderately common	1–12
P. reidi Fowler & Bean, 1928	5–7, 10, 13, 14, 16, 20, 24, 25, 27, 29, 32, 37–39, 42, 46, 47	Moderately common	12–70
P. simsiang Bleeker, 1856	2–4, 8, 11, 18, 20, 43	Occasional.	0–10

Species	Site Records	Abundance	Depth (m)
P. smithi Fowler and Bean, 1928	2–5, 11, 17–19, 21, 22, 24, 29, 32, 35, 41–45	Moderately common.	2–14
P. taeniometopon Bleeker, 1852	4	Rare, but its primary habitat not adequately surveyed.	0–5
P. tripunctatus Cuvier, 1830	29	Rare, except locally common at one site.	0–3
P. vaiuli Jordan & Seale, 1906	39, 46	Rare, several seen, but only at two sites.	3–45
Premnas biaculeatus (Bloch, 1790)	3–7, 9, 10, 12, 20, 24, 25, 30, 31, 35, 42, 43, 45	Moderately common	1–6
Pristotis obtusirostris (Günther, 1862)	44	Rare, only one juvenile seen.	5–80
Stegastes fasciolatus (Ogilby, 1889)	8, 13, 26, 28, 46	Occasional.	0–5
S. lividus (Bloch & Schneider, 1801)	5, 18, 20, 25, 30, 32, 33, 35	Occasional.	1–5
S. nigricans (Lacepède, 1802)	2, 35, 30, 35	Occasional.	1–12
Labridae			
Anampses caeruleopunctatus Rüppell, 1828	28	Rare, only one pair seen	2–30
A. geographicus Valenciennes, 1840	26–28, 33, 37, 38, 45	Occasional.	5–25
A. melanurus Bleeker, 1857	27, 40	Rare, only two seen below depth of 20 m.	12–40
A. meleagrides Valenciennes, 1840	1, 6, 9, 10, 16, 25–28, 33, 37, 39, 40, 47	Occasional.	4–60
A. twistii Bleeker, 1856	7, 10, 13, 25, 37, 39, 45, 47	Occasional.	2–30
Bodianus anthioides (Bennett, 1831)	7, 16, 27, 37	Rare.	6–60
B. bilunulatus (Lacepède, 1801)	27	Rare.	10–80
B. bimaculatus Allen, 1973	25, 47	Rare, but locally common in deep water on steep drop-offs at two sites.	30–60
B. diana (Lacepède, 1802)	1, 7, 9, 10, 12–16, 21, 24, 25, 27, 28, 32, 33, 36–41, 45–47	Moderately common.	6–25
B. mesothorax Schneider, 1801	1, 3, 5–7, 9, 10, 12–16, 19, 20, 22–30, 32, 33, 36–40, 46, 47	Moderately common.	5–30
Cheilinus chlorurus (Bloch, 1791)	6, 17, 23, 31, 32, 34–36, 39, 41	Occasional.	2–30
C. fasciatus (Bloch, 1791)	1–10, 12, 15–26, 29, 30, 32, 34–37	Common.	4–40
C. oxycephalus Bleeker, 1853	1, 15, 17, 18, 21, 24, 26, 29, 30, 32–35	Moderately common.	1–40
C. trilobatus Lacepède, 1802	1, 4, 6–9, 15, 17, 20, 25–30, 32, 33, 36–39, 41–44, 46 47	Moderately common	1–30
C. undulatus Rüppell, 1835	1, 12, 20, 28, 37, 47	Occasional.	2–60
Cheilio inermis Forsskål, 1775	2, 4, 5, 23, 28–30, 33, 35–40	Moderately common.	0–3
Choerodon anchorago (Bloch, 1791)	2, 4–6, 8, 11, 18, 20, 22, 23, 29–35, 37, 38, 41, 43, 44	Moderately common.	1–25
C. sp.	32	Rare, but locally common at one site on rubble bottom at depth of 30 m; one female collected with spear.	25–45
C. zamboangae (Seale & Bean, 1907)	33	Rare.	10–40
C. zosterophorus (Bleeker, 1868)	10, 26, 27, 31, 32, 34, 36, 39–43	Occasional.	10–40
Cirrhilabrus aurantidorsalis Allen & Kuiter, 1999	Jan–25	Common at Togean Islands.	5–40

Species	Site Records	Abundance	Depth (m)
C. cyanopleura (Bleeker, 1851)	26–47	Moderately common.	5–30
C. exquisitus Smith, 1957	25, 28, 47	Rare.	6–32
C. lubbocki Randall and Carpenter, 1980	42	Rare.	6–45
C. solorensis Bleeker, 1853	26, 27, 29, 32, 33, 38, 39, 42, 45, 46	Occasional.	5–35
C. tonozukai Allen & Kuiter, 1999	26, 32–34, 36, 46	Occasional.	10–35
Coris batuensis (Bleeker, 1862)	6, 8, 10, 15, 16, 20, 21, 25–47	Moderately common.	3–25
C. dorsomacula Fowler, 1908	27, 34	Rare.	4–25
C. gaimardi (Quoy & Gaimard, 1824)	1, 7, 10, 12–14, 16, 25–34, 36, 39–42, 46, 47	Moderately common.	1–50
C. pictoides Randall & Kuiter, 1982	31–34, 36, 40, 41	Occasional.	8–30
Diproctacanthus xanthurus (Bleeker, 1856)	1–10, 15–17, 19–47	Common.	2–15
Epibulus insidiator (Pallas, 1770)	1, 2, 4–30, 32, 33, 35, 37–40, 42, 45–47	Common.	1–40
Gomphosus varius Lacepède, 1801	1, 2, 6–10, 12–17, 19, 24–26, 28–30, 33, 34, 36–40, 42, 45–47	Common.	1–30
Halichoeres argus (Bloch and Schneider, 1801)	3, 4, 8, 18, 20, 25, 37, 38	Occasional.	0–3
H. chlorocephalus Kuiter and Randall, 1995	4, 5, 29	Rare, only three seen.	20–30
H. chloropterus (Bloch, 1791)	2, 4–6, 8, 11, 15, 17–20, 22, 23, 25, 31, 32, 35, 41, 43, 44	Moderately common.	0–10
H. chrysus Randall, 1980	1, 6, 7, 9, 12–16, 20, 25–28, 32, 34, 36, 38–40, 46, 47	Moderately common.	7–60
H. hartzfeldi Bleeker, 1852	18, 32–34, 36	Occasional.	10–30
H. hortulanus (Lacepède, 1802)	1, 6, 7, 9, 10, 12–17, 19, 20, 23–30, 32–34, 36–42, 45–47	Moderately common.	1–30
H. leucurus (Walbaum, 1792)	1–8, 11, 15–24, 29, 31, 35, 43, 44	Moderately common	1–20
H. margaritaceus (Valenciennes, 1839)	1, 8, 14, 25, 26, 28, 33, 34, 36–42	Occasional.	0–3
H. marginatus (Rüppell, 1835)	6, 8, 13, 16–18, 25, 26, 28, 29, 33, 34, 36, 39, 40, 46	Occasional.	1–30
H. melanurus Bleeker, 1853	1–10, 13–23, 25, 26, 29, 30, 32–47	Common.	2–15
H. melasmopomus Randall, 1980	7, 13, 14, 16, 21, 24, 25, 27, 42	Occasional.	10–55
H. miniatus (Valenciennes, 1839)	33	Rare, but locally common in shallows at one site.	0–3
H. nebulosus Valenciennes, 1839	27	Rare.	1–40
H. ornatissimus (Garrett, 1863)	37	Rare, only two seen.	5–25
H. papilionaceus (Valenciennes, 1839)	3, 4, 17, 18, 20, 32, 33	Occasional.	0–5
H. podostigma (Bleeker, 1854)	25, 29, 30, 33–35	Occasional.	1–8
H. prosopeion Bleeker, 1853	1, 3, 5, 7, 9, 10, 12–17, 19–34, 36–47	Common.	5–40
H. richmondi Fowler and Bean, 1928	10, 25, 27, 36, 39, 40, 42	Occasional.	1–12
H. scapularis Bennett, 1832	10, 25, 27, 36, 39, 40, 42	Occasional.	1–12
H. solorensis (Bleeker, 1853)	27, 30, 32–36, 41	Occasional.	0–20
H. trimaculatus Griffith, 1834	30, 37, 41	Rare.	0–20

Species	Site Records	Abundance	Depth (m)
Hemigymnus fasciatus Bloch, 1792	1, 7, 10, 26, 28, 30, 37, 47	Occasional.	1–20
H. melapterus Bloch, 1791	1, 3–47	Common.	2–30
Hologymnosus annulatus (Lacepède, 1801)	13, 27–29, 38–46	Occasional.	5–30
H. doliatus Lacepède, 1801	16, 24, 27–30, 33, 36, 41, 46	Occasional.	4–35
Labrichthys unilineatus (Guichenot, 1847)	1, 2, 3–10, 12–27, 29, 30, 32–47	Common.	1–20
Labroides bicolor Fowler and Bean, 1928	1,5–7, 12, 14–22, 24, 25, 27, 29, 30, 32–34, 37–41, 43, 45	Occasional.	2–40
L. dimidiatus (Valenciennes, 1839)	1–9, 11, 13, 15–47	Moderately common.	1–40
L. pectoralis Randall and Springer, 1975	1, 7, 9, 10, 12–15, 17, 19, 21, 23–26, 29, 30, 32, 36, 38, 40, 42, 46, 47	Occasional.	2–28
Labropsis alleni Randall, 1981	5, 7, 9, 10, 17, 19, 21, 23, 25	Occasional.	4–52
L. manabei Schmidt, 1930	9, 17, 24, 27, 33, 35, 38	Occasional.	10–35
Leptojulis cyanopleura (Bleeker, 1853)	1, 3, 33, 36, 39, 40, 43, 46	Occasional.	5–25
Macropharyngodon meleagris (Valenciennes, 1839)	1, 13, 15, 25–28, 30, 33, 34, 36, 39–41, 46, 47	Occasional.	1–30
M. negrosensis Herre, 1932	1, 15, 40, 41, 46	Occasional.	8–30
Novaculichthys macrolepidotus (Bloch, 1791)	37	Generally rare, except locally common in seagrass at one site.	1–6
N. taeniourus (Lacepède, 1802)	1, 8, 13, 36, 38–40, 42, 46	Occasional.	1–14
Oxycheilinus bimaculatus Valennciennes, 1840	1, 2, 6, 33, 34, 36, 38, 39	Occasional.	5–30
O. celebicus Bleeker, 1853	1–6, 8, 10, 15, 17–20, 22–24, 26, 27, 29, 32, 34, 35, 42–45	Moderately common.	3–30
O. diagrammus (Lacepède, 1802)	1, 5, 6, 8, 10, 15, 17, 21, 24–29, 33, 34, 36–41, 45–47	Moderately common.	3–120
O. orientalis (Günther, 1862)	1, 6, 7, 10, 11, 15, 20, 25, 32–34, 38, 40, 42, 45	Moderately common.	15–70
Paracheilinus cyaneus Kuiter & Allen, 1999	7, 9, 12, 13, 14, 32, 44, 45, 46	Occasional.	15–40
P. filamentosus Allen, 1974	16, 32–36, 38, 39, 42, 43, 46	Occasional.	10–50
P. flavianalis Kuiter & Allen, 1999	35	Rare.	10–40
P. togeanensis Kuiter & Allen, 1999	1, 3, 7, 20, 23, 24	Occasional.	15–40
Pseudochelinops ataenia Schultz, 1960	35	Rare, only one seen, but is a cryptic species that seldom exposes itself.	6–25
Pseudocheilinus evanidus Jordan & Evermann, 1902	1, 7, 9, 10, 12–17, 19–21, 24–48, 33, 46, 47	Occasional.	6–40
P. hexataenia (Bleeker, 1857)	1, 2, 6–17, 19–30, 32–34, 36–42, 45–47	Common.	2–35
P. octotaenia Jenkins, 1900	38, 40	Rare.	2–50
Pseudocoris heteroptera (Bleeker, 1857)	36, 37, 40	Rare.	10–30
P. philippina Fowler and Bean, 1928	27, 32, 39	Rare.	10–30

Species	Site Records	Abundance	Depth (m)
P. yamashiroi (Schmidt, 1930)	16, 25, 38	Rare.	10–30
Pseudodax moluccanus (Valenciennes, 1840)	1, 6, 7, 9, 10, 14–17, 25, 27, 28, 37–40, 46, 47	Occasional.	3–40
Pteragogus enneacanthus (Bleeker, 1856)	24, 28–30, 33, 39, 42, 45	Occasional.	2–15
Stethojulis bandanensis (Bleeker, 1851)	6, 7, 10, 13, 16, 18, 25, 27, 28, 33, 36, 37, 39–42, 46	Moderately common.	0–30
S. interrupta (Bleeker)	2, 37, 38	Occasional.	4–25
S. strigiventer (Bennett, 1832)	1, 2, 5, 6, 8, 18, 26, 33, 36, 39	Occasional.	0–6
S. trilineata (Bloch and Schneider, 1801)	1, 2, 5, 6, 8–10, 13, 18, 20, 24, 25, 29, 30, 32, 33, 36, 39, 40, 41, 42, 46, 47	Moderately common.	1–10
Thalassoma amblycephalum (Bleeker, 1856)	1, 6, 7, 13–16, 19, 24–29, 33, 34, 36–39, 42, 45–47	Moderately common.	1–15
T. hardwicke (Bennett, 1828)	1, 2, 5–10, 12, 13, 15–17, 19–30, 32–34, 36–47	Common.	0–15
T. jansenii Bleeker, 1856	13, 26–28, 34, 36–42, 47	Moderately common.	0–15
T. lunare (Linnaeus, 1758)	1, 2, 3–10, 12–30, 32–47	Abundant	1–30
T. trilobatum (Lacepède, 1801)	28	Rare, several seen at one site.	0–5
Wetmorella albofasciata Schultz & Marshall, 1954	9, 15	Rare, only two seen, but is a cryptic species that seldom exposes itself	6–45
Xyrichtys pentadactylus (Linnaeus, 1758)	13	Rare, only five seen.	4–30
Scaridae			
Bolbometopon muricatum (Valenciennes, 1840)	7, 13, 21, 29, 30, 35	Occasional.	1–30
Cetoscarus bicolor (Rüppell, 1828)	1, 7, 10, 11, 15–18, 20, 21, 23–25, 27, 31, 35, 36, 42, 44, 46	Moderately common.	1–30
Chlorurus bleekeri (de Beaufort, 1940)	1–26, 28–47	Common.	2–30
C. microrhinos (Bleeker, 1854)	1, 5–10, 13–26, 29–32, 36, 47	Moderately common.	3–25
C. sordidus (Forsskål, 1775)	1, 6, 7, 9, 10, 15, 16, 24, 25, 28, 32–34, 36–42, 45, 46	Moderately common.	1–25
Hipposcarus longiceps (Bleeker, 1862)	13, 17, 18, 20, 21, 25, 34–39, 41–44	Moderately common.	5–40
Leptoscarus vaigiensis (Quoy & Gaimard, 1824)	37, 38	Rare, but locally common in seagrass at two sites.	1–15
Scarus atropectoralis Schultz, 1958	21, 25, 47	Occasional.	8–30
S. chameleon Choat & Randall, 1986)	1, 28, 29, 34, 36, 40, 41	Occasional.	3–15
S. dimidiatus Bleeker, 1859	1–10, 13–26, 28–30, 32–40, 42, 44–47	Common.	1–15
S. festivus Valenciennes, 1839	45	Rare.	3–30
S. flavipectoralis Schultz, 1958	1, 3–10, 15, 17–26, 29, 31–47	Common.	8–40
S. forsteni (Bleeker, 1861)	16, 25, 28, 46	Occasional.	3–30
S. frenatus Lacepède, 1802	9, 25, 28, 29, 34, 36, 38, 40, 47	Occasional.	3–25
S. ghobban Forsskål, 1775	2, 4, 6, 10, 20, 25, 31, 33, 37, 47	Occasional.	3–30
S. globiceps Valenciennes, 1840	9, 24	Rare, only two seen.	2–15
S. niger Forsskål, 1775	1, 2, 5–7, 9, 10, 12–30, 32–34, 36–40, 43, 45–47	Common.	2–20
S. oviceps Valenciennes, 1839	26, 30, 33, 34, 40, 46, 47	Occasional.	1–12

Species	Site Records	Abundance	Depth (m)
S. prasiognathos Valenciennes, 1839	16, 18, 21, 45–47	Occasional.	2–15
S. psittacus Forsskål, 1775	15, 16, 28, 31, 34, 36–38, 40–42, 46, 47	Occasional.	4–25
S. quoyi Valenciennes, 1840	2–6, 9, 11, 13, 15, 17, 19–25, 28, 29, 31, 32, 37, 39–41, 43–45	Moderately common.	4–18
S. rivulatus Valenciennes, 1840	32, 34, 40, 41, 43, 46	Occasional.	5–20
S. ubroviolaceus Bleeker, 1849	1, 12, 14, 15, 25–28, 34, 36, 37	Occasional.	1–30
S. schlegeli (Bleeker, 1861)	40, 41, 47	Occasional.	1–45
S. spinus (Kner, 1868)	1, 5–7, 9, 10, 12–16, 19, 21, 25, 27–29, 33, 34, 38–41	Moderately common.	2–18
S. tricolor Bleeker, 1849	12–16, 27, 45–47	Occasional.	8–40
Trichonotidae			
Trichonotus setiger Bloch and Schneider, 1801	31	Rare.	3–15
Pinguipedidae			
Parapercis clathrata Ogilby, 1911	1, 26–28, 30, 31, 34, 36–41, 46	Moderately common.	3–50
P. cylindrica (Bloch, 1792)	29, 30, 32, 33, 35, 37, 43	Ocassional	0–20
P. schauinslandi (Steindachner, 1900)	10, 39	Rare.	15–50
P. sp. 1	3, 8, 18, 23, 44	Occasional.	15–50
P. sp. 2	32, 33	Rare.	5–35
P. sp. 3	3, 32	Rare.	10–40
P. tetracantha (Lacepède, 1800)	1, 12, 14–16, 25–28, 32, 46	Occasional.	8–40
Pholidichthyidae			
Pholidichthys leucotaenia Bleeker, 1856	15, 21, 29, 32, 35	Occasional.	1–40
Tripterygiidae			
Enneapterygius tutuilae Jordan & Seale, 1906	18, 23, 28, 33, 42	Occasional.	0–32
Helcogramma striata Hansen, 1986	7, 16, 27, 33, 34, 40, 41, 46	Occasional.	1–20
H. trigloides (Bleeker, 1858)	28	Rare.	0–5
Ucla xenogrammus Holleman, 1993	19, 22–24	Occasional.	2–40
Blenniidae			
Aspidontus taeniatus Quoy & Gaimard, 1834	1, 24, 27, 46	Rare.	1–20
Atrosalarias fuscus (Rüppell, 1835)	1–12, 15, 17–26, 29, 30, 33, 35, 39, 42–45	Moderately common.	1–12
Blenniella chrysospilos (Bleeker, 1857)	28	Rare.	0–3

Species	Site Records	Abundance	Depth (m)
Blenniid sp.	28	Rare.	0–3
Cirripectes castaneus Valenciennes, 1836	9, 21, 24–26, 28, 45, 46	Occasional.	1–5
C. polyzona (Bleeker, 1868)	28	Rare.	0–3
Ecsenius bathi Springer, 1988	27, 33, 47	Rare.	3–20
E. bicolor (Day, 1888)	10, 12–14, 16, 21, 24, 25, 32	Occasional.	2–25
E. bimaculatus Springer, 1971	29, 39–44	Occasional.	5–15
E. lineatus Klausewitz, 1962	32, 36, 44	Rare.	10–25
E. lividinalis Chapman & Schultz, 1952	30–32, 35, 41, 43, 44	Occasional.	2–15
E. midas Starck, 1969	40	Rare.	5–30
E. namiyei (Jordan and Evermann, 1903)	26–28, 32, 39–45	Moderately common	5–30
E. pictus McKinney and Springer, 1976	1, 7, 13, 46, 47	Occasional.	10–40
E. sp. 1	1, 2, 5–8, 10, 15–20, 22–24	Common.	2–10
E. sp. 2	2–8, 10, 17–24, 43–45	Moderately common.	1–20
E. sp. 3	47	Rare, about 10–15 individuals seen in 6–10 m depth on upper edge of steep drop-off at one site.	1–15
E. trilineatus Springer, 1972	47	Rare, only one seen.	2–20
E. yaeyamensis (Aoyagi, 1954)	26, 28, 30, 33, 36, 46	Occasional.	1–15
Meiacanthus atrodorsalis (Günther, 1877)	1, 3, 5, 8, 10, 16, 17, 19, 23, 27, 38, 41–44, 46	Occasional.	1–30
M. ditrema Smith–Vaniz, 1976	29	Rare.	2–15
M. grammistes (Valenciennes, 1836)	5, 6, 8, 11, 23, 25, 27, 29–44	Moderately common.	1–20
M. sp.	1, 3, 11, 18, 22	Occasional.	1–5
M. vicinus Smith–Vaniz, 1987	30, 31, 35, 44	Occasional.	1–10
Nannosalarias nativitatus (Regan, 1909)	28	Rare.	1–12
Paraliticus amboinensis (Bleeker, 1857)	3	Rare, only one seen.	0–2
Petroscirtes breviceps (Valenciennes, 1836)	31	Rare.	0–5
Plagiotremus laudandus (Whitley, 1961)	15, 37, 42	Occasional.	2–35
P. rhinorhynchus (Bleeker, 1852)	1, 2, 5, 7, 10, 13, 16, 19, 24–27, 30, 33–44, 47	Moderately common.	1–40
P. tapeinosoma (Bleeker, 1857)	30, 33, 40	Rare.	1–25
Salarias fasciatus (Bloch, 1786)	18, 25, 30, 32, 33, 37, 38, 41, 43	Occasional.	0–8
S. patzneri Bath, 1992	6, 18–20, 25	Occasional.	1–8
S. ramosus Bath, 1992	31, 41	Rare, only two seen.	2–12
S. segmentatus Bath and Randall, 1991	2–5, 8, 11, 17–19, 22, 23	Occasional.	2–30
Stanulus seychellensis Smith, 1959	28	Rare, only one seen.	0–3

Species	Site Records	Abundance	Depth (m)
Callionymidae			
Callionymus ennactis Bleeker, 1879	5, 29, 41	Occasional.	2–20
C. pleurostictus Fricke, 1992	37	One specimen collected with rotenone.	0–8
Synchiropus morrisoni Schultz, 1960	36	Rare.	12–40
S. splendidus (Herre, 1927)	44	Rare, one specimen seen among live coral branches.	1–18
Eleotridae			
Calumia profunda Larson & Hoese, 1980	14	Rare.	8–40
Gobiidae			
Amblyeleotris guttata (Fowler, 1938)	6, 10, 16, 17, 21, 26	Occasional.	10–35
A. gymnocephalus (Bleeker, 1853)	13	Rare.	5–20
A. latifasciata Polunin and Lubbock, 1979	1	Rare.	15–30
A. periophthalma (Bleeker, 1853)	6–8, 10, 29, 31, 39, 41, 44	Occasional.	8–15
A. randalli Hoese and Steene, 1978	21	Rare.	12–50
A. sp. 1	12	Rare.	10–20
A. steinitzi (Klausewitz, 1974)	3, 6, 8, 10, 16, 20, 22, 25, 31, 38, 41, 44	Occasional.	6–30
A. wheeleri (Polunin & Lubbock, 1977)	14, 27, 34, 36, 39	Occasional.	6–30
A. yanoi Aonuma and Yoshino, 1996	1	Rare.	10–20
Amblygobius buanensis (Herre, 1927)	4	Rare.	0–4
A. decussatus (Bleeker, 1855)	2–5, 8, 10, 11, 15, 17–23, 31, 43, 44	Moderately common.	3–20
A. nocturnus (Herre, 1945)	11, 43	Rare.	3–30
A. phalaena (Valenciennes, 1837)	29, 35	Rare.	1–20
A. rainfordi (Whitley, 1940)	1, 3, 4–7, 10, 15, 17, 19, 21–24, 29, 31, 32, 39, 41–45	Moderately common.	5–25
Asterropteryx bipunctatus Allen and Munday, 1996	19	Rare.	15–40
A. striatus Allen and Munday, 1996	1, 6, 7, 13, 14, 30	Occasional.	5–20
Bryaninops amplus Larson, 1985	8, 16	Occasional, but difficult to detect unless one is specifically searching for it.	5–25
B. loki Larson, 1985	8	Occasional, but difficult to detect unless one is specifically searching for it.	6–45
B. natans Larson, 1986	1, 3, 41	Rare.	6–27
Callogobius sp. 1	44	One specimen collected with rotenone	2–15
Cryptocentrus cinctus (Herre, 1936)	41	Rare.	2–15
C. fasciatus (Playfair and Günther, 1867)	11	Rare.	1–5
C. sp.	8, 11	Rare.	10–25
C. strigilliceps (Jordan and Seale, 1906)	11	Rare.	1–6

Species	Site Records	Abundance	Depth (m)
Crenogobiops crocineus Smith, 1959	19	Rare.	3–10
C. feroculus Lubbock & Polunin, 1977	8, 17, 18, 23, 32	Occasional.	2–15
C. tangaroai Lubbock and Polunin, 1977	42, 46	Rare.	4–40
Eviota albolineata Jewett and Lachner, 1983	15, 16, 20	Occasional.	2–18
E. bifasciata Lachner and Karanella, 1980	1–25, 35, 41, 43–45	Abundant	5–25
E. guttata Lachner and Karanella, 1978	41	Rare.	1–12
E. lachdeberei Giltay, 1933	20	Rare.	3–20
E. latifasciata Jewett & Lachner, 1983	15, 21	Rare.	2–25
E. melasma Lachner & Karanella, 1980	23	Rare.	2–12
E. nigriventris Giltay, 1933	17–19, 31	Occasional.	4–20
E. pellucida Larson, 1976	1–10, 13, 15–25, 33, 41, 43–45	Moderately common	3–20
E. prasites Jordan & Seale, 1906	5, 7, 16, 19, 22, 23	Occasional.	3–15
E. sebreei Jordan & Seale, 1906	1, 4, 5, 10, 23, 24, 34, 36, 42	Occasional.	3–20
E. sp. 1	28	Rare.	3–15
Exyrias bellisimus (Smith, 1959)	2, 11	Rare.	1–25
Fusigobius longispinus Goren, 1978	31	Rare.	8–25
F. neophytus (Günther, 1877)	2, 3, 31, 40	Occasional.	2–15
F. signipinnis Hoese and Obika, 1988	2, 3, 5–7, 10, 11, 15, 17, 20–24, 31, 32, 34, 37–44, 47	Moderately common.	10–30
F. sp. 1	31, 40, 43	Occasional.	5–25
Gnatholepis cauerensis (Bleeker, 1853)	7, 10, 21, 23, 33, 38, 42, 43	Occasional.	1–45
G. scapulostigma Herre, 1953	6, 7, 10, 15, 20, 21, 23, 24, 32–34, 38, 39, 47	Moderately common.	3–30
Gobiid sp.	5	Rare.	4–20
Gobiodon okinawae Sawada, Arai & Abe, 1973	2, 4, 5, 19, 21, 23, 31, 32, 35, 43, 45	Occasional.	2–12
G. quinquestrigatus (Valenciennes, 1837)	7	Rare.	2–12
G. spilophthalmus Fowler, 1944	2	Rare.	1–15
Istigobius decoratus (Herre, 1927)	4	Rare.	1–18
I. ornatus (Rüppell, 1830)	3	Rare.	0–5
I. rigilius (Herre, 1953)	5, 16, 17, 20, 25, 27, 37, 40	Occasional.	0–30
Macrodontogobius wilburi Herre, 1936	2, 5, 23, 31, 32, 43	Occasional.	2–15
Oplopomus oplopomus (Valenciennes, 1837)	8	Rare.	2–25
Paragobiodon echinocephalus Rüppell, 1830	22	Rare.	1–15
Phyllogobius platycephalops (Smith, 1964)	4	Rare.	2–18
Pleurosicya labiata (Weber, 1913)	45	Rare.	3–30
P. micheli Fourmanoir, 1971	38	Rare.	15–40
P. mossambica Smith, 1959	25, 41	Rare.	1–15

Species	Site Records	Abundance	Depth (m)
P. muscarum (Jordan and Seale, 1906)	45, 47	Rare.	2–30
Priolepis cincta (Regan, 1908)	43	Rare.	5–40
Signigobius biocellatus Hoese & Allen, 1977	4, 6, 8, 18, 20, 43	Occasional.	2–30
Stonogobiops nematodes Hoese and Randall, 1982	10	Rare.	10–25
Trimma benjamini Winterbottom, 1996	16	Rare.	6–25
T. caesiura (Jordan and Seale, 1906)	3	Rare.	2–12
T. griffithsi Winterbottom, 1984	5, 16, 19, 23–25	Occasional.	20–40
T. okinawae (Aoyagi, 1949)	45	Several collected with rotenone.	4–25
T. rubromaculata Allen and Munday, 1995	21, 23	Rare.	20–35
T. sp. 1	2, 4, 5, 33	Occasional.	4–20
T. sp. 2	47	Several collected with rotenone	10–25
T. sp. 3	43	Several collected with rotenone	10–25
T. sp. 4	14–16, 21, 47	Several collected with rotenone	8–40
T. sp. 5	15, 19, 21	Several collected with rotenone	8–25
T. sp. 6	13, 15, 21, 47	Several collected with rotenone	8–30
T. sp. 7	2, 4, 5, 23, 31	Occasional.	20–40
T. striata (Herre, 1945)	23, 43–45	Occasional.	2–25
T. taylori (Lobel, 1979)	15	Several collected with rotenone	15–50
T. tevegae Cohen & Davis, 1969	5, 7, 9, 10, 12, 13, 15–17, 19–21, 23–25, 47	Moderately common.	8–45
Valenciennea bella Hoese and Larson, 1994	34	Rare.	18–40
V. helsdingenii (Bleeker, 1858)	33, 39	Rare.	1–30
V. muralis (Valenciennes, 1837)	20	Rare.	1–15
V. puellaris (Tomiyama, 1936)	1, 6, 10, 13, 16, 26, 27, 29, 34, 44	Occasional.	2–30
V. randalli Hoese and Larson, 1994	2, 4, 23	Rare.	8–30
V. sexguttata (Valenciennes, 1837)	4, 10, 13, 16, 18, 20, 31, 37	Occasional.	1–10
V. strigata (Broussonet, 1782)	16, 28	Rare.	1–25

Microdesmidae

Species	Site Records	Abundance	Depth (m)
Aioliops megastigma Rennis and Hoese, 1987	4, 11, 44	Rare.	1–15
Gunnelichthys curiosus Dawson, 1968	1, 33	Rare.	2–30
Nemateleotris decora Randall and Allen, 1973	14–16, 25	Occasional.	28–70
N. magnifica Fowler, 1938	1, 6, 7, 10, 12–14, 16, 21, 24, 25, 42, 46, 47	Moderately common.	6–61
Parioglossus formosus (Smith, 1931)	3, 20	Rare.	0–2
Ptereleotris evides (Jordan & Hubbs, 1925)	6, 8, 13, 16–18, 23, 26–30, 34, 36–39, 44, 46	Moderately common.	2–15

Species	Site Records	Abundance	Depth (m)
P. heteroptera (Bleeker, 1855)	6, 14, 16, 27, 31, 38, 39, 46	Occasional.	6–50
P. monoptera Randall and Hoese, 1985	7	Rare.	5–20
Ephippidae			
Platax boersi Bleeker, 1852			
P. orbicularis (Forsskål, 1775)			
P. pinnatus (Linnaeus, 1758)			
P. teira (Forsskål, 1775)			
Siganidae			
Siganus argenteus (Quoy & Gaimard, 1824)	1, 25, 26, 30, 32, 35–38, 41	Occasional.	1–30
S. canaliculatus (Park, 1797)	3, 4, 31, 37, 38	Occasional.	0–50
S. corallinus (Valenciennes, 1835)	1, 2, 5–7, 9, 10, 14, 17, 24–27, 29, 32, 33, 36, 38, 41, 45	Occasional.	4–25
S. guttatus (Bloch, 1787)	3	Rare, only one seen.	1–15
S. lineatus (Linnaeus, 1835)	4, 13, 16, 18, 20	Occasional.	1–25
S. punctatissimus Fowler & Bean, 1929	13, 15–17, 19–21, 24, 25, 27–29, 38, 39, 41, 45	Moderately common.	3–30
S. punctatus (Forster, 1801)	2–7, 9, 10, 35	Occasional.	1–40
S. virgatus (Valenciennes, 1835)	1–6, 8–10, 13, 15–20, 22, 23, 26, 35	Moderately common	2–25
S. vulpinus (Schlegel & Müller, 1844)	1–10, 12–30, 32–35, 37, 38, 40–47	Common.	1–30
Zanclidae			
Zanclus cornutus Linnaeus, 1758	Jan–47	Common, one of few fishes seen at every site	1–180
Acanthuridae			
Acanthurus auranticavus Randall, 1956	41	Rare.	3–15
A. blochi Valenciennes, 1835	1, 6, 9, 11, 12, 38, 43, 46	Occasional.	3–20
A. fowleri de Beaufort, 1951	2, 5, 9, 10, 12, 14–16, 24, 25, 29, 33, 47	Occasional.	10–30
A. leucocheilus Herre, 1927	13, 28, 33, 37, 38, 40, 41, 45, 46	Occasional.	5–20
A. lineatus (Linnaeus, 1758)	1, 6, 8, 13, 14, 16–19, 21, 24–28, 30, 33, 34, 38, 40, 41, 45–47	Moderately common.	1–15
A. maculiceps (Ahl, 1923)	13, 14, 28, 29	Occasional.	3–20
A. mata (Cuvier, 1829)	3, 7, 9, 10, 12, 14, 16, 24, 25, 27–34, 36–40	Moderately common.	5–30
A. nigricans (Linnaeus, 1758)	1, 7, 13, 15, 24, 25, 28, 36–39, 46, 47	Occasional.	3–65
A. nigricaudus Duncker and Mohr, 1929	1, 7, 10, 16, 18, 20, 21, 23, 25, 26	Occasional.	3–30
A. nigrofuscus (Forsskål, 1775)	24, 26–28, 34, 36, 38, 39, 45, 47	Occasional.	2–20
A. nubilus (Fowler and Bean, 1929)	1, 7, 9, 10, 12, 14–16, 19, 20, 21, 24, 25	Occasional.	10–30
A. olivaceus Bloch & Schneider, 1801	1, 6, 13, 14, 26, 28, 30, 33, 37, 38, 41, 42, 46	Occasional.	5–45

Species	Site Records	Abundance	Depth (m)
A. pyroferus Kittlitz, 1834	1, 6–10, 12–21, 23–47	Common.	4–60
A. thompsoni (Fowler, 1923)	1, 7, 9, 10, 12–16, 19, 21, 24, 25, 30, 33, 37–39, 41, 42, 46, 47	Moderately common.	4–75
A. triostegus (Linnaeus, 1758)	28, 33, 40, 47	Occasional.	0–90
A. xanthopterus Valenciennes, 1835	12–14, 16, 20, 31, 38, 41, 43	Occasional.	3–90
Ctenochaetus binotatus Randall, 1955	2, 3, 5, 6, 17,20, 26, 27, 29–47	Common.	10–55
C. striatus (Quoy & Gaimard, 1824)	1–3, 4–30, 32–47	Abundant	2–30
C. strigosus (Bennett, 1828)	1, 3, 6–8, 12, 15–17, 19, 20, 24, 25, 29, 32, 33, 37–39	Moderately common.	3–25
C. tominiensis Randall, 1955	3–5, 7, 9–24	Occasional.	5–40
Naso brevirostris (Valenciennes, 1835)	7, 25	Rare.	4–50
N. caeruleacauda Randall, 1994	37, 39, 42, 46, 47	Occasional.	15–50
N. hexacanthus (Bleeker, 1855)	12–16, 21, 24, 25, 27, 28, 37, 42	Occasional.	6–140
N. lituratus (Bloch & Schneider, 1801)	1–10, 12–25, 28–30, 33, 36, 37, 39, 42, 45–47	Moderately common.	5–90
N. lopezi Herre, 1927	12, 13, 15, 24, 37	Occasional.	6–70
N. minor (Smith, 1966)	27	Rare.	10–50
N. thymnoides (Valenciennes, 1835)	1, 6, 24, 25, 27, 28, 36, 39	Occasional.	8–50
N. tuberosus Lacepède, 1801	36	Rare, only two seen.	3–20
N. unicornis Forsskål, 1775	2, 5, 25, 29, 33, 44, 45, 47	Occasional.	4–80
N. vlamingii Valenciennes, 1835	4, 9, 12–16, 23, 36	Occasional.	4–50
Paracanthurus hepatus (Linnaeus, 1758)	27, 28, 36–38, 40, 42	Occasional.	2–40
Zebrasoma scopas Cuvier, 1829	1–10, 12–34, 36–47	Abundant	1–60
Z. veliferum Bloch, 1797	1, 2, 4–11, 13–25, 29, 30, 35, 37, 38, 43, 44, 46	Common.	4–30

Sphyraenidae

Species	Site Records	Abundance	Depth (m)
Sphyraena barracuda (Walbaum, 1792)			
S. flavicauda Rüppell, 1838			
S. jello Cuvier, 1829			

Scombridae

Species	Site Records	Abundance	Depth (m)
Grammatorcynus bilineatus (Quoy & Gaimard, 1824)	16, 47	Rare, although a small school seen at Site 16.	10–40
Gymnosarda unicolor (Rüppell, 1836)	25	Rare, only one seen	5–100
Rastrelliger kanagurta (Cuvier, 1816)	35, 38, 41, 42, 45	Occasional schools encountered.	0–30
Scomberomorus commerson (Lacepède, 1800)	14, 27, 41	Rare, only three large adults seen.	0–30

Bothidae

Species	Site Records	Abundance	Depth (m)
Bothus pantherinus (Rüppell, 1830)	37	One juvenile collected with rotenone.	2–110

Species	Site Records	Abundance	Depth (m)
Balistidae			
Balistapus undulattus (Park, 1797)	1–10, 12–30, 32, 34, 36–47	Common.	3–50
Balistoides conspicillum (Bloch & Schneider, 1801)	15, 16, 20, 21, 25, 27, 28, 34, 37, 42	Occasional.	10–50
B. viridescens (Bloch & Schneider, 1801)	10, 12, 13, 15, 16, 20, 23, 40, 46	Occasional; a large female guarding eggs seen at Site 46.	5–45
Melichthys niger (Bloch, 1786)	46	Rare, but locally common in 2–5 m depth at edge of slope at one site; an infrequently seen	
M. vidua (Solander, 1844)	1, 6, 7, 9, 10, 12–17, 19–21, 24, 25, 27, 28, 36–40, 42, 45–47	Common.	3–60
Odonus niger Rüppell, 1836	6, 7, 9, 10, 12–14, 16, 20, 21, 24–26, 28, 29, 32, 33, 38–40, 42, 46, 47	Abundant, literally thousands seen at Site 26	3–40
Pseudobalistes flavimarginatus (Rüppell, 1828)	3, 13, 18, 20, 31, 41, 44	Occasional.	2–50
P. fuscus (Bloch & Schneider, 1801)	3, 20, 36	Rare, one juvenile and two adults seen.	8–50
Rhinecanthus verrucosus (Linnaeus, 1758)	6, 18, 20, 28–30, 33, 35, 38, 41	Occasional.	0–3
Sufflamen bursa (Bloch & Schneider, 1801)	1, 3, 5–30, 32–34, 36–40, 42–44, 46	Common.	3–90
S. chrysoptera (Bloch & Schneider, 1801)	1, 6, 10, 13, 14, 18, 20, 22, 25, 26–28, 30, 31, 33, 34, 36, 38, 39, 41, 42, 46	Moderately common.	1–35
Monacanthidae			
Acreichthys tomentosus (Linnaeus, 1758)	31	Rare, only one seen.	0–3
Aluterus scriptus (Osbeck, 1765)	42	Rare, only one seen.	2–80
Amanses scopas Cuvier, 1829	1, 7, 9, 10, 12, 15, 17, 24–26, 28, 33, 36, 42, 46	Occasional.	3–20
Cantherines fronticinctus (Günther, 1866)	28, 40, 46	Rare, only three seen	2–40
C. pardalis (Rüppell, 1866)	9, 10, 25, 27, 28, 30, 32–34, 36–40, 42, 46	Occasional.	2–20
Oxymonacanthus longirostris Bloch & Schneider, 1801	25, 33, 38, 40, 45	Occasional.	1–30
Paraluteres prionurus (Bleeker, 1851)	31, 32, 40, 46	Rare, only six individuals seen	2–25
Pervagor nigrolineattus (Herre, 1927)	3–5, 8, 31	Rare, only five seen, but difficult to detect due to small size and cryptic habits.	2–15
Ostraciidae			
Lactoria fornasini (Bianconi, 1846)	32, 37	Rare, only two seen.	1–15
Ostracion cubicus Linnaeus, 1758	27, 32–34, 37, 39–41, 44, 45	Occasional.	1–40
O. meleagris Shaw, 1796	7, 14, 26, 30, 39, 42, 45, 46	Occasional.	2–30
O. solorensis Bleeker, 1853	9, 13, 25, 29, 42, 46	Occasional.	1–20

Species	Site Records	Abundance	Depth (m)
Tetraodontidae			
Arothron caeruleopunctatus Matsuura, 1994	1, 10, 16, 20, 24	Occasional.	5–30
A. hispidus (Linnaeus, 1758)	38	Rare, only one seen.	1–50
A. manilensis (de Proce, 1822)	30	Rare, only one seen	1–20
A. nigropunctatus (Bloch & Schneider, 1801)	1, 7, 8, 10, 15–17, 21, 22, 25, 27, 29, 30, 33, 38, 39, 42, 45	Occasional.	2–35
A. stellatus (Schneider, 1801)	13, 27	Rare, only two seen.	3–58
Canthigaster bennetti (Bleeker, 1854)	1, 6, 25, 37, 41	Occasional, except extraordinarily abundant at Sites 1 and 6 where thousands of juveniles were endountered forming schools of more than 200 individuals in each.	1–10
C. compressa Procé, 1822	31, 37	Rare, except locally common among stubby black coral in 28 m depth at Site 31.	1–30
C. coronata (Vaillant & Sauvage, 1875)	36	Rare, one pair seen.	6–40
C. ocellicincta Allen and Randall, 1977	9	Rare, one seen in cave on steep drop–off	15–40
C. solandri (Richardson, 1844)	2, 3, 5, 8, 10, 13, 17, 21–23, 25, 29, 36, 38, 41–44, 46	Moderately common	1–36
C. valentini (Bleeker, 1853)	14, 25, 29–31, 33, 35–37, 39–44, 46, 47	Moderately common; particularly common at Sites 35 and 40.	3–55
Diodontidae			
Diodon hystrix Linnaeus, 1758	18, 25	Rare, only two seen.	2–50
D. liturosus Shaw, 1804	36, 37, 39	Rare, only three seen.	3–35

Appendix 6

Dominant species and percentage of occurrence of target and indicator fishes

See text for explanation of groups A, B, and C. As a guide to interpreting this table, *Aethaloperca rogaa* (the first species) was observed at 4.00 percent of all Togean sites and the numbers counted represented 2.14 percent of all serranids seen in the Togeans, and 0.28 percent of all solitary fishes (Group A) that were counted in the Togeans. % Occ. = % Occurance

	Togean Island			Peninsula			Banggai Island		
	% Occ.	% Family	% Group	% Occ.	% Family	% Group	% Occ.	% Family	% Group
TARGET FISHES—GROUP A									
Serranidae									
Aethaloperca rogaa	4.00	2.14	0.28						
Anyperodon leucogrammicus	20.00	2.14	0.28	33.33	3.33	0.50	15.79	3.13	0.31
Cephalopolis argus	40.00	11.54	1.52	33.33	13.33	2.00	15.79	3.13	0.31
C. cyanostigma	44.00	12.82	1.69	33.33	6.67	1.00	15.79	5.21	0.52
C. boenack	12.00	3.42	0.45				5.26	8.33	0.83
C. leopardus	36.00	5.56	0.73				5.26	1.04	0.10
C. microprion	48.00	14.96	1.98	66.67	13.33	2.00	15.79	9.38	0.93
C. miniata	32.00	10.68	1.41	33.33	30.00	4.50	42.11	27.08	2.69
C. polleni	4.00	0.43	0.06				5.26	3.13	0.31
C. sexmaculatus	24.00	8.12	1.07				10.53	2.08	0.21
C. sonnerati	8.00	1.28	0.17						
C. spiloparaea	12.00	4.27	0.56						
C. urodeta	32.00	14.10	1.86	33.33	13.33	2.00	15.79	8.33	0.83
Epinephelus fasciatus							15.79	9.38	0.93
E. maculayus				33.33	3.33	0.50			
E. merra	12.00	1.71	0.23	33.33	6.67	1.00	15.79	5.21	0.52
Epinephelus sp.	8.00	0.85	0.11						
Gracila albomarginata	4.00	0.43	0.06	33.33	3.33	0.50			
Plectropomus albimaculatus	8.00	1.28	0.17						
P. areolatus	4.00	0.43	0.0						
P. laevis							5.26	1.04	0.10

	Togean Island			Peninsula			Banggai Island		
	% Occ.	% Family	% Group	% Occ.	% Family	% Group	% Occ.	% Family	% Group
P. leopardus	16.00	1.71	0.23				5.26	1.04	0.10
P. maculatus	12.00	1.28	0.17				5.26	3.13	0.31
P. oligocanthus	4.00	0.43	0.06				15.79	5.21	0.52
Variola louti	4.00	0.43	0.06	33.33	6.67	1.00	15.79	3.13	0.31
V. albomarginatus							5.26	1.04	0.10
Lutjanidae									
Aprion viriscens							5.26	1.53	0.21
Aphareus furca	8.00	0.24	0.11						
Lutjanus biguttatus	24.00	14.91	6.95	33.33	19.05	6.00	5.26	2.29	0.31
L. bohar	32.00	3.27	1.52	33.33	4.76	1.50	15.79	9.16	1.24
L. boutton	8.00	17.58	8.19						
L. carponotatus	24.00	3.27	1.52				26.32	16.03	2.17
L. decussatus	92.00	16.24	7.57	66.67	7.94	2.50	36.84	9.92	1.34
L. fulfivlamma	8.00	2.67	1.24	33.33	19.05	6.00	10.53	7.63	1.03
L. fulvus	40.00	9.09	4.23	66.67	6.35	2.00	47.37	18.32	2.48
L. gibbus	8.00	1.58	0.73	33.33	3.17	1.00	15.79	5.34	0.72
L. johni							5.26	0.76	0.10
L. kasmira	16.00	13.70	6.38	33.33	12.70	4.00	5.26	7.63	1.03
Lutjanus lutjanus				33.33	23.81	7.50			
L. monostigma	8.00	0.24	0.11				15.79	4.58	0.62
Macolor macularis	52.00	13.82	6.44	33.33	3.17	1.00	15.79	11.45	1.55
M. niger	12.00	0.36	0.17				15.79	4.58	0.62
Paracaesio sordidus	4.00	3.03	1.41						
Symphorus nematopterus							5.26	0.76	0.10
Haemulidae									
Plectrohynchus chaetodontoides	44.00	57.63	1.92				26.32	8.70	0.62
P. gaterinoides	32.00	15.25	0.51				42.11	42.03	3.00
P. goldmanni	12.00	25.42	0.85				10.53	33.33	2.38
P. lineatus	4.00	1.69	0.06						
P. orientalis				66.67	41.67	2.50	15.79	7.25	0.52
P. polytaenia				33.33	58.33	3.50	15.79	8.70	0.62
Lethrinidae									
Lethrinus erythropterus	44.00	20.00	1.81				31.58	11.76	0.83
L. harak	12.00	5.63	0.51				15.79	17.65	1.24
L. nebulosus							5.26	2.94	0.21
L. obessus							5.26	7.35	0.52
L. obsoletus							5.26	1.47	0.10
L. olivaceus							5.26	7.35	0.52
Lethrinus sp.							5.26	7.35	0.52
Gnatodentex aurolineatus	4.00	3.13	0.28	33.33	37.50	3.00			
Monotaxis granduculus	76.00	71.25	6.44	66.67	62.50	5.00	57.89	44.12	3.10

	Togean Island			Peninsula			Banggai Island		
	% Occ.	% Family	% Group	% Occ.	% Family	% Group	% Occ.	% Family	% Group
Nemipteridae									
Scolopsis bilineatus	20.00	4.98	0.62	100.0	8.60	9.50	73.68	19.37	7.64
S. ciliata	4.00	0.45	0.06				10.53	1.83	0.72
S. lineata	12.00	17.19	2.15				15.79	11.78	4.65
S. margaritifer	72.00	66.06	8.24	66.67	1.81	2.00	47.37	18.32	7.23
S. xenochrous				33.33	1.36	1.50			
Pentapodus caninus	16.00	6.79	0.85	66.67	8.60	9.50	68.42	45.03	17.77
P. emeryi							15.79	1.05	0.41
P. maculatus							5.26	0.26	0.10
P. trivittatus	20.00	4.52	0.56				10.53	2.36	0.93
Mullidae									
Parupeneus bifasciatus	68.00	44.12	6.78	66.67	32.35	5.50	47.37	18.02	4.13
P. barberinus	60.00	20.22	3.11	66.67	17.65	3.00	68.42	21.17	4.86
P. cyclostomus				33.33	5.88	1.00	26.32	4.95	1.14
P. multifasciatus	68.00	18.01	2.77	100.0	20.59	3.50	73.68	41.44	9.50
P. pleuristigma							10.53	1.35	0.31
Mulloidichthys flavolineatus	16.00	16.91	2.60	33.33	17.65	3.00	5.26	9.01	2.07
M. vanicolensis				33.33	5.88	1.00	10.53	3.15	0.72
Upeneus tragula	4.00	0.74	0.11				10.53	0.90	0.21
TARGET FISHES—GROUP B									
Caesionidae									
Caesio caerulaurea	8.00	2.34	1.72				10.53	4.69	2.62
C. cuning	4.00	1.17	0.86				10.53	2.22	1.24
C. diagramma	4.00	0.39	0.29				10.53	4.22	2.36
C. lunaris	12.00	7.89	5.80				5.26	2.35	1.31
C. pisang	4.00	1.80	1.32						
C. teres	76.00	44.90	32.98				21.05	19.24	10.73
C. trilineata	4.00	7.82	5.74						
C. xanthonota							10.53	4.22	2.36
Pterocaesio diagramma	4.00	0.47	0.34				5.26	0.63	0.35
P. marri							5.26	1.56	0.87
P. randalli	20.00	16.80	12.34				10.53	17.21	9.60
P. pisang				33.33	61.54	43.8	15.79	19.56	10.91
P. tessellata							5.26	15.64	8.73
P. tile	12.00	14.21	10.56	66.67	38.46	27.41	15.79	8.45	4.71
Pterocaesio sp.	4.00	1.50	1.11						
Siganidae									
Siganus canaliculatus							21.05	16.22	0.84
S. coralinus	4.00	0.49	0.01				10.53	1.01	0.05
S. doliatus	48.00	21.46	0.49	33.33	8.00	0.44	26.32	7.09	0.37

	Togean Island			Peninsula			Banggai Island		
	% Occ.	% Family	% Group	% Occ.	% Family	% Group	% Occ.	% Family	% Group
Siganus fuscensens				33.33	60.00	3.29	10.53	23.65	1.22
S. guttatus	8.00	4.88	0.11						
S. linetaus							5.26	1.69	0.09
S. puelloides	8.00	1.95	0.04	33.33	4.00	0.22	15.79	2.03	0.10
S. puellus	48.00	13.17	0.30	66.67	16.00	0.88	26.32	4.05	0.21
S. punctatus	4.00	0.98	0.02						
S. punctatissimus	24.00	8.29	0.19				47.37	11.15	0.58
S. virgatus	24.00	7.32	0.17				15.79	2.03	0.10
S. vulpinus	76.00	40.98	0.93	33.33	12.00	0.66	84.21	29.39	1.52
Siganus sp.	16.00	0.49	0.01				5.26	1.69	0.09
Acanthuridae									
Acanthurus blochii	20.00	2.28	0.53				21.05	1.61	0.63
A. grammoptilus	8.00	0.38	0.09				10.53	1.16	0.45
A. leucocheilus	4.00	0.19	0.04	100.0	15.09	3.51	26.32	1.21	0.47
A. lineatus	24.00	2.28	0.53	33.33	5.66	1.32	26.32	1.52	0.59
A. lopezi							5.26	0.27	0.10
A. mata							10.53	2.46	0.96
A. nigricans	16.00	0.62	0.14	66.67	2.83	0.66	21.05	0.67	0.26
A. nigricauda							5.26	0.22	0.09
A. nubilis	12.00	1.62	0.38				10.53	0.89	0.35
A. olivaceus	4.00	0.14	0.03				5.26	0.27	0.10
A. pyroferus	52.00	7.32	1.71	100.0	18.87	4.39	84.21	7.10	2.77
A. thompsoni	36.00	16.26	3.80				36.84	7.24	2.83
A. triostegus	4.00	1.43	0.33						
Naso brevirostris	4.00	0.10	0.02				10.53	2.90	1.13
N. hexacanthus	12.00	4.04	0.94				15.79	12.96	5.06
Naso lituratus	88.00	5.71	1.33	33.33	7.55	1.75	47.37	2.37	0.92
N. thynnoides	8.00	1.76	0.41				15.79	7.69	3.00
N. unicornis	4.00	0.05	0.01				5.26	0.67	0.26
N. vlamingii	16.00	0.57	0.13				10.53	0.27	0.10
Naso sp.	4.00	0.05	0.01				5.26	0.09	0.03
Paracanthurus hepatus				33.33	5.66	1.32	21.05	0.71	0.28
Ctenohaetus binotatus	48.00	6.80	1.59				47.37	11.71	4.57
C. striatus	80.00	14.88	3.48	100.0	22.64	5.26	89.47	19.12	7.47
C. tominiensis	68.00	13.98	3.27						
Zebrasoma scopas	80.00	14.74	3.45	100.0	21.70	5.04	73.68	16.18	6.32
Z. veliferum	64.00	4.80	1.12				26.32	0.71	0.28

	Togean Island			Peninsula			Banggai Island		
	% Occ.	% Family	% Group	% Occ.	% Family	% Group	% Occ.	% Family	% Group
TARGET FISHES—GROUP C									
Carangidae									
Caranx bajad	12.00	3.57	1.18				5.26	2.70	0.16
C. ferdau	4.00	1.19	0.39				5.26	2.70	0.16
C. melampygus	32.00	61.90	20.39	33.33	100.0	100.0	10.53	10.81	0.65
C. sem	4.00	1.19	0.39						
C. sexfasciatus	4.00	3.57	1.18						
Caranx sp.							5.26	5.41	0.33
Carangoides bajad	12.00	3.57	1.18						
Carangoides ferdau	4.00	1.19	0.39				5.26	2.70	0.16
Decapterus macarellus							5.26	54.05	3.26
Decapterus sp.	4.00	23.81	7.84						
Elagatis bipinnulata							5.26	16.22	0.98
Trachinotus baelloni							5.26	5.41	0.33
Sphyraenidae									
Sphyraena baraccuda	4.00	0.62	0.39						
Sphyraena pinguis	8.00	49.69	31.37						
S. jello	4.00	49.69	31.37						
Scombridae									
Rastrelliger kanagurta	4.00	100.0	3.92				10.53	99.83	93.80
Gymbosarda unicolor							5.26	0.17	0.16
INDICATOR FISHES									
Chaetodontidae									
Chaetodon adiergastos	4.00	0.12	0.12				15.79	0.41	0.41
C. auriga							47.37	1.47	1.47
C. baronessa	72.00	4.04	4.04	100.0	7.63	7.63	47.37	3.02	3.02
C. benneti	16.00	0.36	0.36	33.33	1.53	1.53	10.53	0.16	0.16
C. burgessi	4.00	0.12	0.12						
C. citrinellus	4.00	0.06	0.06	33.33	2.29	2.29	15.79	0.49	0.49
C. ephippium	36.00	1.09	1.09				10.53	0.33	0.33
C. kleinii	56.00	5.19	5.19	100.0	38.17	38.17	100.0	36.87	36.87
C. lineolatus							10.53	1.39	1.39
C. lunula	24.00	0.72	0.72				15.79	0.57	0.57
C. lunulatus	80.00	7.79	7.79	33.33	3.05	3.05	89.47	7.59	7.59
C. melannotus	12.00	0.36	0.36	33.33	3.05	3.05	57.89	2.45	2.45
C. meyeri	28.00	1.93	1.93	33.33	0.76	0.76			
C. ocellicaudus	40.00	0.84	0.84	33.33	1.53	1.53	21.05	0.41	0.41
C. octofasciatus	40.00	7.60	7.60				10.53	4.24	4.24
C. ornatissimus	48.00	2.23	2.23				10.53	0.33	0.33

	Togean Island			Peninsula			Banggai Island		
	% Occ.	% Family	% Group	% Occ.	% Family	% Group	% Occ.	% Family	% Group
C. oxycephalus	28.00	0.91	0.91				26.32	1.14	1.14
C. punctatofasciatus	68.00	9.23	9.23	33.33	1.53	1.53	36.84	2.20	2.20
C. rafflesi	80.00	3.56	3.56	33.33	1.53	1.53	26.32	1.47	1.47
C. selene				33.33	1.53	1.53	15.79	0.73	0.73
C. semeion	40.00	1.45	1.45				10.53	0.24	0.24
C. speculum	4.00	0.06	0.06				26.32	0.82	0.82
C. trifascialis	28.00	1.33	1.33	100.0	6.11	6.11	47.37	3.67	3.67
C. ulietensis	48.00	2.66	2.66	33.33	1.53	1.53	47.37	1.55	1.55
C. unimaculatus	12.00	2.60	2.60	66.67	4.58	4.58	26.32	1.22	1.22
C. vagabundus	72.00	3.26	3.26	66.67	7.63	7.63	47.37	2.28	2.28
C. xanthurus	8.00	0.24	0.24				26.32	1.14	1.14
Coradion melanopus	44.00	3.02	3.02	100.0	7.63	7.63	63.16	1.96	1.96
Forcipiger longirostris	60.00	5.37	5.37	66.67	3.05	3.05	42.11	1.71	1.71
F. flavissimus							57.89	4.73	4.73
Hemitaurichthys polylepis	28.00	22.93	22.93				15.79	10.44	10.44
Heniochus acuminatus	8.00	2.60	2.60				5.26	0.16	0.16
H. chrysostomus	60.00	2.78	2.78	66.67	3.05	3.05	36.84	1.79	1.79
H. depreutes	4.00	1.21	1.21						
H. singularius	24.00	0.97	0.97				10.53	0.24	0.24
H. varius	64.00	3.38	3.38	66.67	3.82	3.82	73.68	2.77	2.77

Appendix 7

Diversity and abundance of target and indicator fishes at each survey site

The number of species is indicated in column A and number of individuals is in column B.

Site No.	Total Spec. (A)	Total Indiv. (B)	Serranid. (A)	Serranid. (B)	Lutjanidae (A)	Lutjanidae (B)	Haemulid. (A)	Haemulid. (B)	Lethrinidae (A)	Lethrinidae (B)	Nemipterid. (A)	Nemipterid. (B)	Mullidae (A)	Mullidae (B)	Caesionidae (A)	Caesionidae (B)	Siganidae (A)	Siganidae (B)	Acanthurid. (A)	Acanthurid. (B)	Carrangidae (A)	Carrangidae (B)	Sphyraenid. (A)	Sphyraenid. (B)	Scombridae (A)	Scombridae (B)	Chaetodontidae (A)	Chaetodontidae (B)
1.	31	162	8	17	4	42					1	1	2	8			1	2	5	41							10	51
2.	31	155	4	14	4	15	3	3	1	1	2	3	2	4	1	30	4	9	2	31	3	22					5	23
3.	29	292	2	2	5	31	1	1	1	2	3	15	1	3	1	140	3	22	2	12	3	26					7	38
4.	25	276	5	5	2	5			2	2	1	7	3	4	3	160	2	7	1	1	1	8	1	50			4	27
5.	23	394	3	4	4	55			1	3	3	12	2	2	1	275	3	11	1	3							5	29
6.	34	140	1	1	5	18	1	1	2	4	1	7	3	21			4	12	7	37	1	8					10	32
7.	36	408	5	17	5	24			2	5			2	3	1	150	1	2	4	115							15	91
8.	31	174	1	1	1	1			2	3	3	27	5	29	2	10	2	9	6	72			1	1			8	21
9.	44	366	5	7	3	6			2	18	1	2	3	5	2	175	2	3	9	86	1	1					16	63
10.	39	730	4	4	2	6			1	4	3	9	2	8	1	500	3	9	8	92							15	98
11.	24	78	3	10	2	2	2	3	3	3	1	4	1	2	1	22	1	1	3	10							7	21
12.	46	1787	8	38	4	58	2	29	1	2	1		2	17	3	1250	1	2	7	156	1	6					17	229
13.	50	732	7	16	5	177	1	6	1	3		1	2	8	2	195	2	2	11	143							19	182
14.	40	1803	7	23	5	143	1	2	1	3			2	3	3	1475	2	2	5	70	1	2					13	80
15.	41	983	3	10	5	37	1	1	2	7	1	1	1	1	2	680	2	2	10	133	1	1					13	110
16.	43	423	2	2	3	22	1	2	3	73	1	3	1	5	1	50	3	5	12	178	1	2					15	81
17.	35	255	6	19	5	25	2	2			1	2	1	1	1	10	1	2	7	116	1	1					10	77
18.	33	188	3	3	2	10			2	8	3	25	3	11	1	40	3	12	4	18			1	30			11	31
19.	37	567	3	7	1	2	1	1	1	5	1	3	2	4	2	400	4	10	7	98	3	3					12	34
20.	44	606	1	2	3	18	2	2	3	5	2	46	4	56	3	250	2	4	8	97			1	80			15	46
21.	44	510	3	4	1	8			1	4	2	29	2	24	2	150	6	27	12	188	1	3					14	73
22.	27	263	2	3	2	7			1	1	2	10	3	6	1	125	4	11	6	53							6	47
23.	30	336	1	2	4	54	1	1	1		2	10	2	3	1	150	4	18	6	58							9	40
24.	46	453	4	6	4	25	2	2	1		1	2	1	27	3	85	4	16	11	220	1	1			1		15	69
25.	44	308	7	18	3	34	2	3	1	4	1	3	2	17	1	75	1	5	10	75						10	15	64
26.	35	269	5	11	6	48	1	2	2	11	3	24	1	2	1	75	4	22	5	39							7	35
27.	37	396	6	19	3	12	1	7	1	5	3	14	4	19	1	200	7		7	56							11	64

Appendix 8

List of species caught by fishers of the Togean and Banggai Islands, Sulawesi, Indonesia

Abbreviations as follows: T = Togean; B = Banggai; t = Togean dialect; b = Bobongko dialect; s = Saluan dialect.

FISHES

Family and Species	Local Name	Location	Notes
Terraponidae		T/B	
Pelates sexlineatus	Bangkatayu (t)		caught using net, string and hook,
Terapon jarbua			bubu
Scolopsis bilineata	Tongtong (s)		fresh consumption or salted
Holocentridae			
Myripristis murdjan	Sogo (s)	T/B	caught using net, string and hook,
M. kuntee	Sogo		bubu
M. hexagona	Sogo		fresh consumption or salted
Sargocentron spiniferum			
S. rubrum			
Haemulidae			
Plectorhinchus polytaenia	Bibiga (t)	T/B	caught using net, string and hook,
P. gibbosus	Bibiga (t)		bubu and dynamite
P. lineatus	Bibiga (t)		fresh consumption or salted
P. orientalis	Bibiga (t)		
P. obscurus	Bibiga (t)		
Priacanthidae			
Priacanthus blochii	Silo-silo	T	
Latjunidae		T/B	
Lutjanus kasmira			caught using net, string and hook,
L. carponotatus			bubu and dynamite
L. decussatus			fresh consumption or salted
L. boutton			
L. gibbus	Dappa (t)		
L. vitta	Bulunga (s)		
L. fulvus	Aragan lappa (s)		
L. ehrenburgi	Baba (t)		
L. fulviflamma			
L. bohar	Tingalu (t)		
L. rivulatus	Jina		
Aprion virescen	Butu		
Macolor macularis	Bolekaluku (s)		

Family and Species	Local Name	Location	Notes
Caesionidae			
Caesio cuning	Lolosi (t)	T/B	caught using net, string and hook,
C. teres	Lolosi		bubu and dynamite
C. xanthonota	Lolosi		fresh consumption or salted
C. lunaris	Bapua (s)		
Pterocaesio randalli	Londou		
P. pisang	Lolosi cabe		
P. marri			
P. chrysozona			
Lethrinidae			
Gnathodentex aurolineatus	Tolotok (t)	T/B	caught using net, string and hook and
Letrinus harak	Katamba (t)		bubu
L. ornatus	Katamba (t)		fresh consumption or salted
L. erythracanthus	Katamba (t)		
L. nebulosus	Tingkolopo		
L. olivaceous			
L. obsoletus			
Monotaxis grandoculis			
Mullidae			
Mulloidichthys vanicolensis	Lumut	T/B	caught using net, string and hook and
M. flavolineatus	Lumut		bubu
Parupeneus bifasciatus	Lumut		fresh consumption or salted
P. barberinus	Lumut		
P. indicus	Lumut		
Scatophagidae		T	
Scatophagus argus	Bete-bete (t)		caught using net, string and hook, bubu and dynamite fresh consumption or salted
Pomacanthidae		T	
Pomacanthus annularis			caught using net, string and hook,
P. imperator			bubu and dynamite
P. semicircularis			fresh consumption or salted
Pomacentridae		T/B	
Amblyglyphidodon curacao			caught using net, string and hook and
Abudefduf lorenzi			bubu
A. sexfasciatus			fresh consumption or salted
A. bangalensis			
Labridae		T/B	
Cheilinus fasciatus	Ikan kea		caught using net, string and hook,
C. undulatus	Mamin/napoleon		bubu, dynamite and traditional poison
Oxycheilinus diagrammus			live fish collection, fresh consumption
O. celebicus			or salted
Pseudocheilinus octotaenia			Napoleon (wrasse) has high economic
Thalasoma lunare			value and have become endangered
T. trilobatum			due to over fishing

Family and Species	Local Name	Location	Notes
Carangidae		T/B	
Caranx sem	Babara		caught using net, string and hook,
C.sexfasciatus	Botuan		bubu and dynamite
C.melampygus	Mosidung		fresh consumption or salted
Carangoides bajad	Dorini		
C. ferdau	Babara		
Seriola dumerili	Lamali		
Elagatis bipinnulata	Uro-uro		
Scoberoides lysan	Lali		
Sphyraenidae		T/B	
Sphyraena barracuda	Baracuda		caught using net, string and hook,
S. jello	Loli-loli		bubu and dynamite
			fresh consumption or salted
Apogonidae			caught using very fine net for
Pterapogon kauderni	Capung Banggai	B	ornamental fish

OTHER BIOTA

Family and Species	Local Name	Location	Notes
Holothuria		T/B	fresh consumption, dried, high
Holothuridae	Teripang pasir		economic value
Holothuria scabra	Teripang getah		
H. edulis	Teripang coklat		
Bohadschia. marmorata	Teripang batu		
Actinopyga. lecanora			
Crustacea		T/B	fresh consumption, high economic
Palinuridae			value
Panulirus femoristriga	Lobster mutiara		
P. versicolor	Lobster pasir		
Coenobitidae			fresh consumption, endangered
Birgus latro	Ketam kenari		
Reptilia		T/B	fresh consumption or decoration,
Cheloniidae	Penyu hijau		endangered
Chelonia mydas	Penyu sisik		
	Penyu		

Coastline of Banggai Island

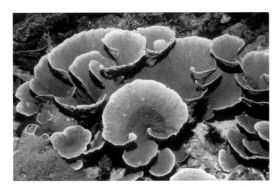

Hard corals *(Leptoseris yabei)*, Togean Islands

Blue damselfish *(Chrysiptera cyanea)*, Togean Islands

Banggai cardinalfish *(Pterapogan kauderni)*.

New species of damselfish *(Ambllypomacentrus clarus)*.

Fisherman in the Banggai Islands.

Hard coral polyps *(Tubastraea* sp.), Togean Islands.

RAP charter boat, *Serenade*.

Pulau Dondola (Site 25).

Anemone crab *(Neopetrolisthes ohshimai)*,
Banggai Islands.

1998 Banggai-Togean Marine RAP Science Team.
From left to right: D. Fenner, K. Anwar, G. Allen, S. Yusuf, T. Werner, and La Tanda.

All photographs by Gerald R. Allen

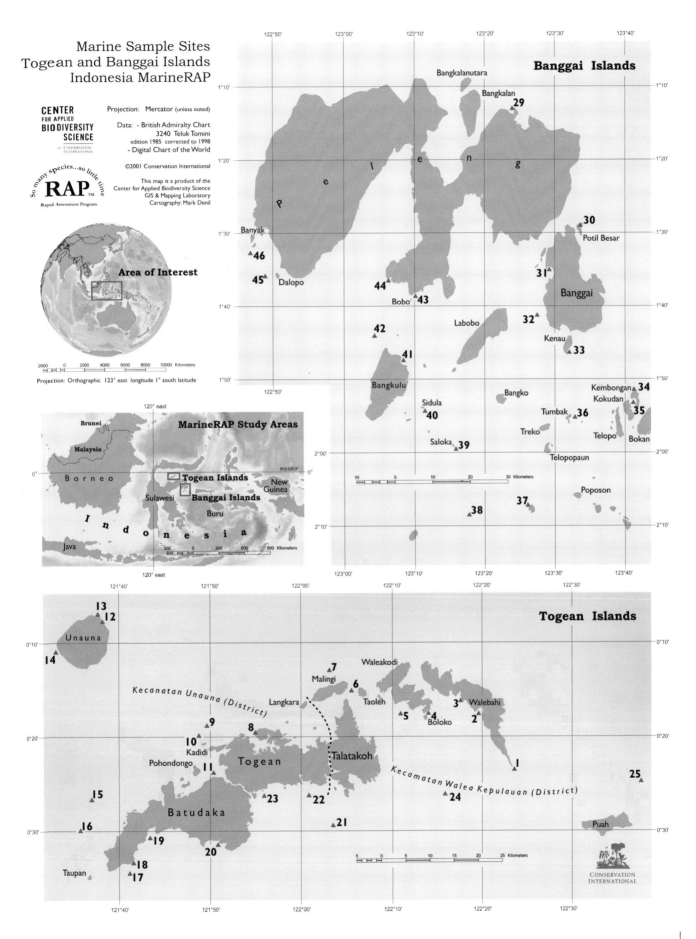

Marine Sample Sites
Togean and Banggai Islands
Indonesia MarineRAP

CENTER
FOR APPLIED
BIODIVERSITY
SCIENCE
AT CONSERVATION
INTERNATIONAL

Projection: Mercator (unless noted)

Data: - British Admiralty Chart
3240 Teluk Tomini
edition 1985 corrected to 1998
- Digital Chart of the World

©2001 Conservation International

This map is a product of the
Center for Applied Biodiversity Science
GIS & Mapping Laboratory
Cartography: Mark Denil

So many species...so little time
RAP™
Rapid Assessment Program

Area of Interest

2000 0 2000 4000 6000 8000 10000 Kilometers

Projection: Orthographic 123° east longitude 1° south latitude

Banggai Islands

Bangkalanutara
Bangkalan
29
30
Potil Besar
Banyak
46
45 Dalopo
31
Banggai
44
Bobo **43**
32
Kenau
Labobo
33
42
41
Bangkulu
Sidula
40
Bangko
Kembongan **34**
Kokudan
Tumbak **36**
35
Treko
Telopo
Bokan
Saloka **39**
Telopopaun
37
Poposon
38

10 0 10 20 30 Kilometers

P   e   l   e   n   g

MarineRAP Study Areas

Brunei
Malaysia
Borneo
Togean Islands
New Guinea
Sulawesi
Banggai Islands
Buru
I n d o n e s i a
Java
equator

300 0 300 600 900 Kilometers

Togean Islands

13
12
Unauna
14
7
Waleakodi
Malingi
6
Taoleh
3 Walebahi
Kecamatan Unauna (District)
Langkara
5
4 **2**
Boloko
9
8
10
Kadidi
Pohondongo
11
Togean
Talatakoh
1
25
15
23
22
Kecamatan Walea Kepulauan (District)
24
Batudaka
21
Puah
16
19
20
18
Taupan
17

5 0 5 10 15 20 25 Kilometers

CONSERVATION
INTERNATIONAL